Concept Mapping as an Assess[ment] Tool for Conceptual Understanding in Mathematics

This book investigates the practicability and effectiveness of the concept map as a tool for assessing students' conceptual understanding in mathematics.

The author first introduces concept mapping and then employs it to investigate students' conceptual understanding of four different mathematical topics. Alongside traditional scoring methods, she adopts Social Network Analysis, a new technique, to interpret student-constructed concept maps, which reveals fresh insights into the graphic features of the concept map and into how students connect mathematical concepts. By comparing two traditional school tests with the concept map, she examines its concurrent validity and discusses its strengths and drawbacks from the viewpoint of assessing conceptual understanding. With self-designed questionnaires, interviews, and open-ended writing tasks, she also investigates students and teachers' attitudes toward concept mapping and describes the implications these findings may have for concept mapping's use in school and for further research on the topic.

Scholars and postgraduate students of mathematics education and teachers interested in concept mapping or assessing conceptual understanding in classroom settings will find this book an informative, inspiring, and overall valuable addition to their libraries.

Haiyue Jin obtained her PhD from the National Institute of Education at Nanyang Technological University in Singapore and is currently a lecturer at the School of Education Science at Nanjing Normal University in China. Her research interests include mathematics assessment, learning strategies in mathematics, and primary and secondary mathematics teaching.

Concept Mapping as an Assessment Tool for Conceptual Understanding in Mathematics

Haiyue Jin

R Routledge
Taylor & Francis Group

LONDON AND NEW YORK

This book is published with financial support from 'Research-oriented General Primary Teacher Training Mode under Synergy Mechanism', a project of Excellent Teacher Training Reform of Jiangsu Province (Project No: Teacher Su [2015] 19).

First published 2022
by Routledge
4 Park Square, Milton Park, Abingdon, Oxon OX14 4RN

and by Routledge
605 Third Avenue, New York, NY 10158

Routledge is an imprint of the Taylor & Francis Group, an informa business

British Library Cataloguing in Publication Data
A catalogue record for this book is available from the British Library

Library of Congress Cataloging in Publication Data
A catalog record for this book has been requested

ISBN: 978-1-032-21643-0 (hbk)
ISBN: 978-1-032-21645-4 (pbk)
ISBN: 978-1-003-26937-3 (ebk)

DOI: 10.4324/9781003269373

Typeset in Times New Roman
by Apex CoVantage, LLC

Contents

Figures

Tables

Preface

This book is based on my doctoral dissertation completed at National Institute of Education, Nanyang Technological University, Singapore.

Over the decades, psychological and educational research on mathematics learning has firmly established the importance of conceptual understanding in our ability to use knowledge flexibly and apply what we learn appropriately in different settings. A variety of techniques have been applied in an attempt to measure levels of conceptual understanding in mathematics. However, researchers have long known that these techniques, including commonly used achievement tests, are highly limited and tend to yield only fragmented information about learners' true understanding of the topics. Accordingly, there is a clear need to explore more direct and informative measurements of conceptual understanding. Concept mapping, a technique initially developed by Novak and Gowin, is widely recommended for that purpose but is not widely used in mathematics education.

This book investigates the feasibility of concept mapping as an assessment of conceptual understanding in mathematics. It is intended to supplement previous studies and address gaps in research and practice in the following aspects. First, this book introduces a concept mapping training programme for students and addresses its effectiveness with three groups of students in the preliminary, pilot, and main studies. The issue of student training has not been carefully considered in previous studies on concept mapping. Such training is necessary for students if they are to construct informative, evaluable concept maps. Training is equally important for educators who are interested in concept mapping but new to the practice and unsure how to adopt it for use in assessment. Second, while previous studies on concept mapping focus on a single topic when investigating the students' conceptual understanding, four different mathematical topics—two in algebra (algebraic expressions and equations) and two in geometry (triangles and quadrilaterals)—are explored here to address possible topic-level differences. Third, in addition to traditional scoring methods, I apply a novel Social Network Analysis-based technique for assessing student-constructed concept maps at the whole-class level. Through this collective approach to analysis, educators can gain fresh insights into the graphic features of the student-constructed concept maps and the connections they perceive among concepts. Fourth, students' performance on concept mapping tasks and their performance on two traditional school testing tasks on the same topics are compared and the concurrent validity of concept mapping is thereby examined. Strengths and drawbacks of concept mapping as a technique for assessing conceptual understanding are also discussed. Fifth, students', prospective teachers', and in-service teachers' attitudes toward concept mapping are examined through self-designed questionnaires along with interviews and open-ended writing tasks. The attitudes of such groups are likely to influence whether concept mapping achieves popularization in educational settings. Based on this, I explore some of the implications of the key findings in terms of concept mapping for school use and suggest avenues for further research.

This book is intended as both a guide to mapping mathematical concepts and interpreting mathematical concept maps in their specific context as an assessment tool and a catalyst for further research on concept mapping as a technique for building and assessing conceptual understanding in mathematics.

Finally, I would like to take this opportunity to thank Professor Wong Khoon Yoong, my PhD supervisor, for his inspiring guidance and continuous encouragement; Prof. Lee Peng Yee, for his unfailing support; my participants, for their time and cooperation; and many scholars whose work has enriched my intellectual life and classroom praxis.

Nanjing, China

1 Introduction

Concept Mapping for Mathematics

Background

Over recent decades, psychological and educational research on mathematics learning has firmly established the importance of conceptual understanding in terms of our ability to use knowledge flexibly and apply what we learn appropriately in different settings (e.g. Bransford et al., 1999; National Council of Teachers of Mathematics (NCTM), 2000). Educators, researchers, and curriculum designers alike (Afamasaga-Fuata'I, 2008; Kilpatrick et al., 2001; Ministry of Education, Singapore, 2019; NCTM, 2000) have acknowledged this importance and contributed to explorations of the field. Moreover, substantial efforts have been made to measure levels of conceptual understanding (e.g. Niemi, 1996a; Webb and Romberg, 1992; White and Gunstone, 1992; Jones et al., 2019; Ceran and Ates, 2020). However, researchers have long known that, as Niemi (1996a, p. 351) puts it, 'commonly used achievement tests provide, at best, only indirect and highly limited information on students' conceptual understanding'. Accordingly, exploration of more direct and informative means of measuring learners' conceptual understanding in mathematics is needed.

Conceptual understanding is by nature a cognitive construct grounded in various and multiple theories, such as Schema Theory (Rumelhart, 1980; Rumelhart and Ortony, 1977) and Piaget's (1977) Equilibration of Cognitive Structure Theory. Hiebert and Lefevre's (1986) description of conceptual knowledge remains highly influential and forms a basis for most contemporary work:

> Conceptual knowledge is characterized most clearly as knowledge that is rich in relationships. It can be thought of as a connected web of knowledge, a network in which the linking relationships are as prominent as the discrete pieces of information. Relationships pervade the individual facts and propositions so that all pieces of information are linked to some network.
>
> (pp. 3–4)

Accordingly, the extent of not only our knowledge of relevant concepts but also of the relationships between them determines the extent of our conceptual

DOI: 10.4324/9781003269373-1

knowledge. A technique that appears to be relevant to the features of conceptual knowledge is Novak and Gowin's (1984) concept mapping.

Concept map is defined as a two-dimensional graphical representation of knowledge. Concepts, referred to as 'nodes', are linked with labelled arrows to denote relations between nodes in a pair (Novak and Gowin, 1984). Although concept mapping has been used extensively in assessing conceptual understanding in science education (Novak, 1990; Wallace and Mintzes, 1990) and is considered effective, the practice remains comparatively rare in mathematics education. As mathematics learning and science learning involve many of the same psychological and epistemological properties, it is worth exploring whether concept mapping can be likewise effective adapted to mathematics (Novak, 2006) and thereby recommended for more extensive applications. This book aims to introduce the application of concept mapping as an assessment of students' conceptual understanding of mathematics.

Toward Deeper Assessment of Conceptual Understanding

In many countries, school education is exam-oriented (e.g. Tan, 2006; Zhang et al., 2004), and although contemporary mathematics curricula emphasise development of learners' conceptual understanding, commonly used school measures primarily emphasise problem-solving proficiency. Lack of relevant and effective assessment tools and strategies that directly address students' conceptual understanding makes implementing such curricular demands difficult and hinders timely intervention when students' understanding is in need of adjustment. This book is a response in particular to the need for effective assessment techniques and tool for measuring students' conceptual understanding in mathematics.

When exploring novel assessment methods, researchers should keep in mind fundamental elements of assessment that ensure that meaningful inferences are drawn. Pellegrino et al. (2001) identified three elements of effective assessment, namely, cognition, observation, and interpretation, also known as the Assessment Triangle. Ruiz-Primo and Shavelson (1996) characterised assessments, particularly concept mapping assessments, in terms of assessment task, students' responses, and scoring of responses. As these two models are complementary, in the present study they are combined to form the following categories, assessment task, student response, and interpretation. The assessment task element involves checking that the task assigned is backed by solid theories or beliefs indicating that it can represent student knowledge in a subject domain. The student response element involves whether the examiner can observe and infer students' performance of the construct being assessed from the responses provided. Interpretation of responses replaces Ruiz-Primo and Shavelson's (1996) scoring because interpreting a concept map is much more than scoring. This element involves checking that the interpretation method selected is valid and reliable for drawing inferences from the students' responses obtained.

For the first element, *assessment task*, broad theories support concept mapping as an effective means of capturing attributes of conceptual understanding.

In cognitive psychology, it is generally agreed that human knowledge is stored in the memory in information packets comprising schemas (Jonassen et al., 1993). When learning occurs, we tend to incorporate new information into our schemas. If the information is contrary to what we already know, we may need to adapt our schemas to better accommodate it. These incorporating and adapting processes are known as assimilation and accommodation, respectively (Piaget, 1977). The balance between the two processes reflects how knowledge develops in the mind. Explorations into concept formation, concept acquisition, and conceptual learning in mathematics (e.g. Sfard, 1991; Skemp, 1987) support the pattern of relations among mathematical concepts and equilibration processes. Skemp argues that to understand a concept is 'to assimilate it into an appropriate schema' (1987, p. 29). From this representational view of understanding (Perkins, 1998), conceptual understanding, which emphasises the understanding of concepts, can be defined as the structured interrelatedness among concepts in a domain. Once conceptual understanding is represented externally, it can be assessed by others. A concept map, with its specific attributes, i.e. nodes, links, linking phrases, and structure, can explicitly represent conceptual understanding. It can serve as a 'window into the mind' (Shavelson et al., 2005, p. 1).

The second element is *student responses*. The two main areas affecting students' responses in concept mapping tasks (CM tasks) are their concept map drawing skills and the quality of their conceptual understanding. Studies on the use of the concept map as an assessment tool usually begin with an introduction to concept map and training on how to construct one (e.g. Ruiz-Primo, Schultz, et al., 2001; Williams, 1994). *Concept mapping* refers to the process of drawing a concept map. Training students on concept mapping is necessary, just as training them on the use of software computer in computer-based test assessment is necessary. Most researchers have found that students can quickly learn to construct concept maps with limited practice (e.g. Freeman, 2004; Ruiz-Primo, 2004; Ruiz-Primo, Schultz, et al., 2001). However, others argue that concept mapping imposes high cognitive demands on students by requiring them to identify important concepts, relationships, and structures within a given domain of knowledge (e.g. Novak and Gowin, 1984; Schau and Mattern, 1997). As such, students may experience initial difficulty constructing concept maps.

These two positions seem somewhat conflicting, and the research has not adequately addressed the optimal amount of training required, taking into account different levels of students. These issues are of concern for the further development and application of concept mapping in assessment. In addition, there is a gap in most studies between students' concept mapping training and the mapping performance criteria used by examiners to gauge their conceptual understanding; that is, examiners (e.g. Edwards and Fraser, 1983; Jin, 2007; Williams, 1998) seldom evaluate students' mapping skills after training but instead proceed directly to assessing students' constructed concept maps, thus assuming that such maps are faithful representations of students' conceptual understanding. These studies assume that participating students have equivalent mapping skills after the same training and that those who draw better concept maps have higher levels of conceptual understanding. These assumptions are questionable because it is quite possible for a student who is experienced in concept mapping but who has a low level of conceptual understanding to construct a higher-level concept map than a student with a high level of conceptual understanding but who is not clear on the concept map construction process. In such cases, it is unfair to assess students' understanding by comparing their concept maps. Skill at mapping and its relationship with the resulting maps should be clarified before going deeper into any study treating students' concept maps as representations of their conceptual understanding.

Once a concept map is completed, interpreting the information it contains becomes the main concern. This is the third element of the Assessment Triangle. Researchers (e.g. McClure et al.'s, 1999) often build scoring systems for concept maps based on features, such as valid nodes, meaningful propositions, and structure. The scores allow examiners to assess conceptual development by directly comparing different students' concept maps or comparing concept maps constructed by the same student at different times. Other researchers (e.g. Afamasaga-Fuata'I's, 2009a, 2009b; Liu and Hinchey, 1996) employ qualitative methods to focus directly on the details of students' performance, e.g. the connections they make that show insight into or misconceptions about the conceptual relationships. Both scoring methods and qualitative methods are valuable for extracting information embedded in student-constructed concept maps.

This though leads naturally to the question of whether a given interpretation of a concept map is valid; that is, whether the information obtained and the inferences drawn fairly represent the quality of students' conceptual understanding. The existing literature, however, does not provide satisfactory answers to such questions, and researchers have found that information collected from concept maps and other measures, e.g. interviews, writing tasks, multiple-choice tests, and the Science Achievement Test (e.g. Edwards and Fraser, 1983; Hoz et al., 1990; Novak, 2005) yields inconsistent conclusions about students' achievement. The inconsistent results may be partially due to students' lack of familiarity with concept mapping (Snead and Snead, 2004), which would leave them inadequately prepared to effectively map their knowledge of a topic. Ruiz-Primo, Schultz, et al. (2001) have also reported that different mapping tasks, e.g. 'fill-in-the-map' and 'construct-a-map' tasks, are not equivalent in their capacity for representing students' knowledge. In addition, evidence to date suggests that concept maps measure different aspects of achievement, compared to other instruments (Anderson and Huang, 1989; Markham et al., 1994; Novak et al., 1983). Different scoring methods may also result in differing conclusions about students' mapping performance (Ruiz-Primo, 2004). Given these inconsistencies, whether a concept map is a valid representation of conceptual understanding should be further examined, with attention to mapping skills and mapping formats.

Based on these considerations, this book is guided by following research questions:

1   Do students need training to construct informative concept maps? What level of training is appropriate and sufficient?
2   What attributes of students' conceptual understanding can be assessed from student-constructed concept maps?

3   What differences among mathematical topics, if any, should be considered in the use of concept mapping as a technique for assessment of students' conceptual understanding?

4   Is concept mapping a valid technique for assessing students' conceptual understanding in mathematics?

5   What are students' and teachers' attitudes toward concept mapping in mathematics?

The main study described in this book began with training on concept mapping to prepare the participating students with basic knowledge about what a concept map is and how to construct one. The effectiveness of the training is examined to ensure that they were prepared with the necessary concept mapping skills. They then constructed concept maps for four different mathematical topics, two algebraic topics and two geometric topics. Together with the mapping task, they completed two types of traditional school tests. These different measurements provide multiple perspectives on their conceptual understanding. Comparison between the measures allows for investigation into the concurrent validity of concept map. Moreover, students' and teachers' attitudes toward concept mapping in mathematics are examined, because their perspectives are important factors to be considered when adopting a new assessment technique. Furthermore, synthesis of students' achievement in school-type measures of conceptual understanding in mathematics with their concept mapping performance, and attitudes toward concept mapping contribute to deeper understanding of concept maps as an assessment tool for students' conceptual understanding of mathematics.

Filling in the Gaps in Concept Mapping in Mathematics Education

In that techniques and tools for evaluating conceptual understanding in mathematics are still lacking despite the need for them, this book is relevant, specifically in the following ways:

Firstly, on the one hand, this book explores concept mapping as an assessment tool for directly evaluating students' conceptual understanding in mathematics. This is especially desirable since more attention to conceptual understanding is currently advocated by curriculum designers and educators. On the other hand, this book argues that the information about students' conceptual understanding revealed by the student-constructed concept maps in concert with the information obtained from the two traditional measurements can cast light on students' instruction needs at the level of specific mathematics topics.

Second, this book works toward filling the gaps in the literature between training on concept mapping and the products of concept mapping. It presents a guide to concept mapping training and describes my efforts to develop a training programme tailored for concept mapping in mathematics, from skills training to evaluation. The results of the investigation suggest the need for further applications for training students in concept mapping and evaluating their concept mapping skills. Insights into the relationship between students' mapping skills and their mapping performance revealed here are intended to inform further exploration and experimentation in the area. Moreover this book addresses scholars' somewhat contradictory views on the validity and reliability of concept mapping as an assessment practice, to the extent that such inconsistencies are due to comparing the work of students' who have unequal familiarity with concept mapping and the skills it requires (Snead and Snead, 2004).

Third, this book investigates the concurrent validity of the concept map as a measure for assessment of four different mathematical topics, by comparing it with two commonly used measures with the same topics. Such investigation adds to findings on the validity of using concept mapping as a technique for assessing students' conceptual understanding. This is desirable because inconsistent findings on validity were reported in the literature.

Fourth, with this book I hope to further enrich the body of literature on students and teachers' attitudes toward concept mapping and provide a reference for teachers who desire to use concept mapping for teaching, learning, and assessment in mathematics. Empirical evidence, in the form of findings on relationships between students' attitudes toward concept mapping and their mapping performance may offer a basis for further exploration of concept mapping in school education in general.

This book specifically caters to the need for additional assessment techniques on conceptual understanding in mathematics and contributes to our current knowledge on using concept mapping for assessment purpose. The explorations and findings carry implications for further application of concept mapping in both research and classroom settings.

Roadmap for the Book

This book comprises eight chapters. Chapter 1 introduces the background, rationale, and purpose of this book and positions it within the research. Chapter 2 clarifies the definition and components of conceptual understanding and reviews techniques for measuring conceptual understanding in mathematics. Chapter 3 includes an overview of concept mapping, concept mapping task formats, interpreting of concept maps, and relevant validity and reliability issues. It also explores the literature on conceptual understanding and the use of concept mapping as an assessment technique in mathematics. Chapter 4 describes the researchers' development and implementation of a concept mapping training programme for mathematics. Various training methods are explained along with sample student-constructed maps. The various training studies I conducted in furtherance of this project make is clear that training is necessary if students are to construct informative, evaluable maps. The effectiveness of the final training programme is carefully examined and improvements to it suggested.

In Chapter 5, the main study is presented. Having undergone the skills training programme (as described in Chapter 4), participants mapped their conceptual understanding of basic algebra and geometric shapes. Multiple methods are applied to analyse the resulting student-constructed maps, including the traditional methods for scoring concept maps proposed by Novak and Gowin and techniques

I adopted from Social Network Analysis. Relations across the mathematical concepts are explicitly addressed alongside examples of well- and poorly constructed concept maps. Insights into students' conceptual understanding revealed through assessment of their concept maps are also presented. In Chapter 6, concept mapping and two traditional type school tests are compared in terms of their effectiveness for capturing students' conceptual understanding in mathematics. The strengths and drawbacks of the concept mapping as an assessment technique are subsequently discussed. Chapter 7 investigates students and teachers' attitudes toward concept mapping through questionnaires, interviews, and writing tasks. Finally, Chapter 8 summarises the main points and reiterates the book's arguments and contributions. Implications for practice in both research and classroom settings are discussed and potential areas for further exploration are highlighted.

2 Conceptual Understanding in Mathematics and Its Measures

Defining Conceptual Understanding in Mathematics

Bell et al. (1983) explain the ideas of *concept* in mathematics as follows:

> The term *concept* has two meanings in common use—one more precise but less important, the other vague but more important. The first meaning distinguishes a *concept* from a *relationship*; here a concept is a 'thing' (e.g. rhombus, group, multiplication) needing, in formal mathematics a definition, while a relationship is expressed by a statement (e.g. 'multiplication of real numbers is commutative'). . . . The second meaning of the term is that implied when one is speaking of 'teaching for concepts' or of 'having the concept of place value'. These are loose expressions. . . . The better term here is 'conceptual structure', meaning a network of concepts and relationships.
>
> (pp. 78–79)

The first meaning of *concept* applies to the individual concept, which is customarily delineated by a name and an accompanying definition. The name and definition apply to a class of items, words, or ideas that share common attributes. For example, the name *rhombus* is accompanied by a definition: 'a quadrilateral with four equal sides.' The meaning of the name rhombus is clarified by the definition. It is a class of quadrilaterals that share the common attribute of having *four equal sides*. In teaching mathematics, examples and nonexamples are usually employed to help students better grasp the definitions. Inclusion of various modes of representation, i.e. numbers, words, other symbols, diagrams, stories, and real things, are recommended, especially in teaching abstract mathematical concepts (Wong, 1999). Connections between representations of an individual concept among different representation modes can yield deeper descriptions and deeper understandings of the concept. Bell et al.'s (1983) second meaning of *concept* is grounded in relationships and is comparatively more abstract than the first meaning. Skemp's idea of *schema* (Skemp, 1987) may help to explicate this second meaning. Skemp does discuss *concept* in illustrating schema, but only in reference

DOI: 10.4324/9781003269373-2

to Bell et al.'s (1983) first meaning. To avoid misunderstanding, unless specified, the term *concept* in this book refers to the first meaning, namely, *concept* as mathematical object.

Skemp (1987) divides concepts into *primary concepts*, those directly derived from experience, and *higher-order concepts*, those built on relationships with or properties of other concepts. He claims that 'each [concept] (except primary concepts) is derived from other concepts and contributes to the formation of yet others' (p. 22). In Figure 2.1, C_1, C_2, and C_3 represent primary concepts which share some common attributes. Based on such attributes, mathematicians can abstract a new idea and name it 'C''. Thus, the abstraction C' is a higher-order concept. The common attributes of C_1, C_2, and C_3 make the newly created concept C' distinguishable from other concepts. From the concept C', together with other concepts, such as C'' and C''', further abstractions of C can be made, and so on and so forth. A hierarchical structure is then formed in the process of generalising the higher-order concepts generated. At the same time, however, Skemp (1987) argued that combining basic concepts randomly is not sufficient for the emergence of a new concept. A new concept is generalisable only when the basic concepts are suitably connected.

For example, in preschool education, *natural numbers*—also called *counting numbers* or simply *numbers*—originated directly from the experience of counting things, beginning with the number *one*. This is what defines it as a primary concept. Conversely, the *prime number* is an example of a higher-order concept. In mathematics, it refers to a natural number that has exactly two distinct natural number factors: one and itself. To arrive at or understand the concept *prime number*, one must first understand the primary concepts *natural numbers* and *divisors*. However, understanding *natural number* and *divisors* is not sufficient. An understanding of the concept *prime number* can be obtained only when the two concepts are appropriately related. This relationship is important in the formation of conceptual understanding of *prime numbers*.

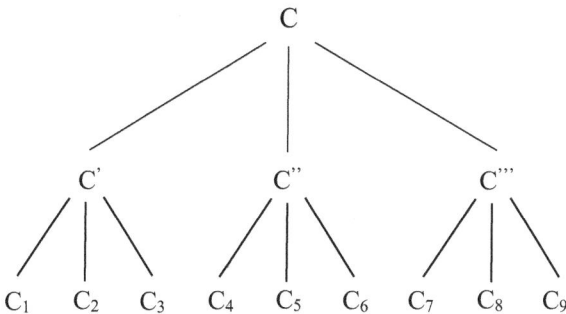

Figure 2.1 A hierarchical structure of mathematical concepts.

Source: Adapted from Skemp's (1987, p. 10) ideas about abstraction and schema in mathematics

Skemp's (1987) idea of schema enriches Bell et al.'s (1983) second meaning of *concept* as 'a network of concepts and relationships' (p. 79) by explaining how such networks are formed. In mathematics, however, conceptual understanding cannot rely solely on definitions and relationships across concepts. It sometimes requires the support of the operation in practice. Some concepts are even defined on the basis of their operation. Consider the same concept *prime number*: when asked the conceptual question of whether 1597 is a prime number, absent external help, e.g. from a computer programme, one can only obtain the answer by carrying out the appropriate operation, checking whether 1597 is divisible by any natural number other than one and itself. Performing this operation can aid our retention of the definition of *prime number*, facilitate its connections with *natural numbers* and *factors* and hence assist us in understanding the concept *prime number*. This may be a reason that the Mathematics Learning Study Committee (MSC) of the National Research Council in the US includes operation in its definition of conceptual understanding: 'comprehension of mathematical concepts, operations, and relations' (Kilpatrick et al., 2001, p. 116). From a practical standpoint, however, this definition is vague because 'comprehension' needs further explication.

To better understand the 'operation' aspect of this definition of conceptual understanding we turn for a moment from the cognitive psychology concept of *acquisition*, which emphasises appropriate schemas or representations, to Perkins (1998) flexible performance view of understanding. He claims that 'to understand a topic means no more or less than to be able to perform flexibly with the topic—to explain, justify, extrapolate, relate, and apply in ways that go beyond knowledge and routine skill' (p. 42). Perkins (1998) further argued that the performance view is superior to the representational view, mainly because representations or mental schemas are sometimes not enough and other times unnecessary for understanding. His justifications are reasonable in some sense, but I would consider it is unwise to adopt only the performance view of conceptual understanding, especially for the teaching and learning of mathematical concepts. Mathematical concepts are abstract; understanding them usually engages students in high levels of cognitive activity. Metaphorically, performance is the output of students' minds. When students can perform flexibly with what they know, we have reason to believe that they have understood the knowledge presented. However, especially at the initial stages of learning, students cannot always perform concept successfully in different cases. Different reasons may account for such failures. Teachers need to address weaknesses in their students' understanding and arrange their teachings accordingly. Performance-based assessment is not sufficient for uncovering the weaknesses or reasons behind such failures. Moreover, it may not be wise for teachers to wait until students fail to perform to uncover flaws in their understanding. In addition to 'output', we should be aware of the 'mechanisms' of students' mental processes. To makes such mechanisms visible and assessable, strategies stemming from the representational view of conceptual understanding are indeed necessary.

In reviewing the assorted perspectives on conceptual understanding described earlier, three major components of conceptual understanding in mathematics emerged:

- Individual concepts.
- Relations among individual concepts.
- Relations among concepts and operations.

Acquiring individual concepts is the first step toward conceptual understanding. In mathematics, this usually includes being able to define the concept, presents examples and nonexamples, and recognise different representations. Only with basic knowledge of individual concepts can meaningful relationships be established. Recognising relationships among concepts is a necessary but also difficult step toward conceptual understanding. It is often the case that students acquire a great deal of knowledge on individual concepts but do not meaningfully connect them (Bereiter and Scardamalia, 1998; Bransford et al., 1999). Operations are components of conceptual understanding since many mathematical concepts are defined by operations, and it is through operations that conceptual understanding can be advanced to a higher level. When concepts and operations are appropriately connected, the relationships established can in turn contribute to understanding individual concepts and fostering further assimilation of other concepts. It is important to note that the third component emphasises the underlying rationale of operation rather than operation itself, that is, the conceptual rather than the procedural aspect of operation.

Techniques for Assessing Conceptual Understanding

In mathematics, students' achievement is often assessed through examination problems, which are typically routine and well practiced. When students succeed in solving this type of problem, teachers tend to assume that they have acquired the underlying concepts and operations. However, solving routine problems successfully does not necessarily imply a proper understanding of relevant concepts (Niemi, 1996a; de Vries et al., 2002). Various assessment techniques have been proposed to measure students' conceptual understanding.

Defining concepts plays an important role in the construction of mathematical concepts (Tirosh, 1999; Morgan, 2005). A student's definition of a given concept is often viewed as a mirror of their understanding of it (Mosvold and Fauskanger, 2013; Panaoura et al., 2017). Accordingly, explicitly defining a mathematical concept is a common conceptual task that students are often asked to perform (Crooks and Alibali, 2014), sometimes with illustrative examples and nonexamples (e.g. Panaoura et al., 2017). However, Vinner (1992) cautioned that ability to define does not guarantee an understanding of the concepts. Students can memorise a definition of a concept by rote or in a meaningful way. For example, in their study with Israeli students, Rasslan and Vinner (1998) found that while most of the participants defined the functions, few could apply the definitions correctly.

To help students understand concepts in-depth, teachers usually employ multiple representations (Wong, 1999) or conceptual variations (Gu et al., 2004). In this way, students' conceptual understanding can then be inferred from multiple representations of the target concepts and conceptual variations. For example, Panasuk (2011) assessed middle-school algebra students' conceptual understanding of a *linear equation with one unknown* via multiple conceptual representations. That study focused on the interpretation, connection, and translation of such representations. Panasuk concluded that a student's ability to present different representation of a given concept or relationship is an indicator that they are advancing from procedural skills to conceptual understanding.

Students' knowledge of the multiple representations of a concept and the relationships among these representations is often the focus of assessment measures featuring an individual concept. Conceptual understanding, however, also requires understanding a concept's relationships with other concepts, including similarities and differences, subsets and supersets, parts and whole, and situation specific operational differences. Products of open-ended tasks requiring students to organise information and express it in their own words (Atlgan et al., 2020) can be used to assess their understanding of these types of relationships (e.g. Taylor, 2008; Price and van Jaarsveld, 2017; Hauer et al., 2020).

A variety of such open-ended tasks have been proposed for the purpose (e.g. Norwood and Carter, 1994; Novak, 2005; Wiburg et al., 2016; Price and van Jaarsveld, 2017). Niemi (1996a), for example, developed multiple measures, including justification and explanation tasks, to assess Grade 5 students' conceptual understanding of *fractions* in Washington state. The justification measure assessed students' ability to justify their problem-solving procedures by explicitly relating symbols to meaningful knowledge. The explanation measure assessed students' ability to explain concepts and draw pictures to support their explanations; specifically, it checked their knowledge of relationships between explicit concepts, principles, situational knowledge, and symbols. Together the measures provide diagnostic information on the level and quality of individual students' understanding of concepts and principles. Interviews are also commonly used to reveal students' deeper thinking and understanding of mathematics. Castellón et al. (2011) investigate the effectiveness of interactive interviews in assessing a group of English language learners' conceptual understanding of *fractions*. After interviewing students twice about their work on a written task, they concluded that interactive interviews are useful in addressing students' understanding of mathematical knowledge and provide opportunities for students to reflect on their mathematical thinking and revise their mathematical hypotheses. Although very powerful in investigating students' deep thinking, such open-ended tasks can address a limited number of knowledge points at a time, and do not present visual representations of students' cognitive structure (Novak, 2005).

In analysing data collected during a 12-year longitudinal study on students' understanding of science concepts, Novak (2005) and his colleagues found it difficult to trace the conceptual development their participants' expressed in interview transcripts. To solve this problem, they invented concept mapping. Concept mapping is a technique for illustrating relationships among concepts through two-dimensional links. They used the technique transform the interview data into concept maps. They found that a 15- to 20-page interview transcript could be easily converted into a one-page concept map without losing essential information about the concepts and propositional meanings. Over the years, concept maps have become known as a powerful tool for specifying relationships between different concepts and tracking changes in students' cognitive structures.

Also in science education, White and Gunstone (1992) discuss eight types of probes for understanding, namely, prediction-observation-explanation, relational diagrams, word association, interviews, concept mapping, drawings, question production, and fortune lines. They used some of these, such as concept mapping, to investigate conceptual understanding. Relational diagrams and word association share many characteristics with concept mapping. To construct relational diagrams, one must draw closed figures to indicate patterns of overlap between classes of objects, events, and abstractions and word association tasks directly elicits the perceived associations for a set of concepts through the propositions she/he makes. However, the more concepts are involved, the more complex a relational diagram becomes and the less easy it may be for others to understand it; thus, the relational diagram may only be adaptable when a limited number of terms is involved. Similarly, word associations can span any number of concepts but cannot indicate the whole structure of the relationships among the concepts. For these reasons, concept mapping may be stronger than relational diagramming and word association tasks since it can be used to illustrate a relatively large number of concepts as will become more apparent in subsequent sections.

Other researchers (e.g. Alonso-Tapia, 2002; Ceran and Ates, 2020; Domin and Bodner, 2012; Mintzes et al., 2000; Savander-Ranne and Kolari, 2003; Schnotz and Preub, 1997; Wilson, 1992) have proposed other powerful assessment techniques, such as V-diagrams, portfolios, classroom observation, dialogue-as-data, representation, and justification, for investigating students' conceptual understanding. The rationale behind most of these techniques involves the assumption that concepts are not isolated and that the meaning of any concept is implicit in the context and in its relations to other concepts (Jonassen et al., 1993). Efforts to validate measures of conceptual understanding in mathematics (e.g. Chiu et al., 2007; Niemi, 1996a) are ongoing, and some such measures are administered in schools, e.g. journal writing and open-ended tasks (Drake and Amspaugh, 1994; Moskal, 2000) with varying degrees of effectiveness. This book contributes to such validation efforts with respect to concept mapping.

Summary

This chapter provides a reference to literature on conceptual understanding and techniques used to assess conceptual understanding in mathematics. As we have seen, the term *conceptual understanding* is commonly used (e.g. Afamasaga-Fuata'I, 2006; Niemi, 1996a) but seldom explicitly defined. Related terms are also used in the literature, such as conceptual knowledge (Hiebert and Lefevre, 1986), relational understanding (Skemp, 1976), and concept image (Vinner, 1983). It is a challenge to identify similarities and differences between *conceptual understanding* and these terms. Instead of becoming entangled with definitions, I identified three key components of conceptual understanding running through the literature: individual concepts, relations among individual concepts, and relations among concepts and operations. In addition to the well-known representational view of conceptual understanding from cognitive psychology, I incorporate Perkins' (1998) performance-based view of understanding. That is, as Perkins posits, our conceptual understanding is more than our possession of the correct mental model for a target concept; we must also understand how to work with that model. The performance view is consistent with the 'operations' element of Kilpatrick et al.'s (2001) definition of conceptual understanding. Consideration of three components of conceptual understanding was practical for designing the instruments described in this book.

When attempting to measure conceptual understanding, traditional school examinations have focused primarily on students' knowledge of individual concepts, formulas, theorems, and/or their operations in solving problems. In order to meaningfully assess their concept understanding, educators must present students with tasks that allow them to directly express their understanding of relationships between concepts and to justify mathematical operations. Open-ended tasks and interviews are very powerful in investigating students' deep thinking. However, they can accommodate only a limited number of knowledge points at a time and, as pointed by Novak (2005), cannot offer visual representations of students' cognitive structures. From this perspective, concept mapping can, at minimum, fill gaps in measuring conceptual understanding by specifying, and even structuring, the relationships between concepts in a certain domain, which is one of the most important components of conceptual understanding.

3 Why Concept Mapping?

Defining the Concept Map

The *concept map* is a graphical representations of knowledge within a particular domain (Novak and Gowin, 1984). As illustrated in Figure 3.1 (Novak and Cañas, 2006), it is a network of nodes and labelled lines. In this model, the nodes, important terms representing concepts, are enclosed in boxes (circles are also often used). Relationships and connections are indicated by connecting lines labelled with *linking words* or *linking phrases* that indicate how the two concepts are related and which include arrows to indicate the direction of the relationship. Linking concepts in this way results in the production of meaningful statements. Such statements are called *propositions*. According to Ruiz-Primo (2004), the proposition is the basic unit of meaning in a concept map and the smallest unit used to judge the validity of the relationship indicated between any two concepts on the map.

A notable characteristics of the concept map specified by Novak and his colleagues (Novak and Cañas, 2006; Novak and Gowin, 1984) is *hierarchical structure*. The concept map must flow from the most inclusive, most general concept positioned at the top of the map to the more specific concepts positioned at lower levels (Novak and Cañas, 2006). While numerous researchers apply this criterion to concept maps (e.g. Liu and Hinchey, 1996; Malone and Dekkers, 1984; Wallace and Mintzes, 1990), others do not impose a hierarchical structure. For example, in the semantic-network tradition, practitioners tends to use spider maps (Ahlberg, 2004; Harnisch et al., 1994). In a spider map, the focus concept is positioned in the centre of a map, and the links develop outward to other related concepts. A generally accepted view is that whether a hierarchical structure is needed depends on the subject domain. If the subject matter is hierarchical, then a hierarchical map is meaningful; otherwise, there is no reason to impose such a structure on one's understanding (White and Gunstone, 1992; cited in Shavelson et al., 1994).

Since its introduction, Novak's definition of concept map has been modified according to various interests and for various purposes, and the term *concept map* is sometimes even used quite differently than as Novak proposed. For example, Kane and Trochim (2007) define concept map as a generic term for any illustration

DOI: 10.4324/9781003269373-3

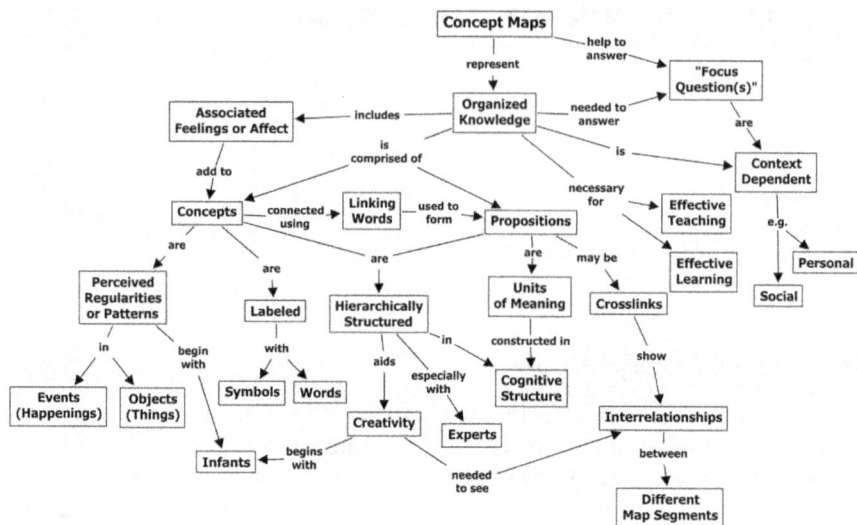

Figure 3.1 A concept map showing the key features of concept maps.

Source: Novak and Cañas, 2006, p. 2

that represents a domain of ideas. According to this definition, a geographical or topographical map could also be considered a concept map.

In this book, *concept map* refers specifically and only to graphic representations comprising *nodes* representing concepts and *labelled lines/arrows* denoting connections among nodes. Nodes are mathematical concepts, examples and non-examples of the concepts, diagrams, symbols, and formulas; the labels or *linking phrases* are verbs or adjective phrases. Although a hierarchical structure might be a better representation of the essential inter-relatedness among mathematical concepts (Skemp, 1987), young learners may have difficulty distinguishing or expressing the hierarchy of abstract concepts (Novak and Gowin, 1984; Schau and Mattern, 1997). On this basis, the student participants described in this study (mainly 8th graders) were encouraged but not required to construct concept maps with hierarchical structures.

Applications of Concept Mapping

Concept mapping have been used most extensively in science education, and it is primarily applied in educational environments as an instructional technique, a learning strategy, and an assessment technique.

Concept Map as an Instructional Tool

Due to its potential for clarifying connections between concepts, for instructional purposes, concept map can be used as advance organisers (Willerman and Mac

Harg, 1991), aids for displaying lesson and course development (e.g. Starr and Krajcik, 1990), and as a means of integrating information (Malone and Dekkers, 1984). It can offer an effective structure for clarifying connections between concepts and ideas into which new concepts and ideas can be integrated.

Classroom teachers regularly encounter students with difficulty understanding science or mathematics concepts that have been carefully taught to them (Malone and Dekkers, 1984). Rote memorisation of conceptual definitions viewed as isolated information by students often takes the blame. Including teacher-constructed concept maps during instruction can efficiently reduce the likelihood that students will memorize material by rote; concept map facilitates teachers' delivery of clear and coherent explanations to students (Novak and Cañas, 2006; Cliburn, 1986). This position is supported by experimental studies. Willerman and Mac Harg (1991) found that eighth-grade students who received instruction using concept maps achieved greater learning of the physical and chemical properties of compounds than their peers who received standard instruction for the unit, as measured by an end-of-unit multiple-choice test. In 1993, Horton and her colleagues published a meta-analysis of research on the effectiveness of concept mapping as an instructional tool for science teaching (Horton et al., 1993). The meta-analytic results suggest that concept mapping generally has positive effects on both student achievement and attitudes.

Concept Mapping as a Learning Method

That concept mapping is an effective strategy that can facilitate meaningful learning has been consistently and persuasively argued (e.g. Heinze-Fry and Novak, 1990; Nesbit and Adesope, 2005). Concept mapping helps students select, organise, and integrate concepts into a knowledge structure. Having a knowledge structure in place facilitates students' concept retention and application of learned concepts in unfamiliar contexts (Novak, 1990). For example, Weinstein and Mayer (1986) describe the use of concept mapping, which they called *networking*, as a strategy for organising complex learning tasks, such as comprehending text. Their review of the literature suggests that students who receive training on context mapping outperform those who do not in retaining and comprehending the main ideas on tests involving materials from unfamiliar biology and physics textbooks. Receiving explicit instruction on reading or constructing concept maps appears to facilitate learners' identification of internal connections among concepts presented in expository materials.

Researchers have invested great effort into studying collaborative concept mapping, which usually involves engaging two or more students working coordinately to learn and construct knowledge (e.g. Gao et al., 2007). Numerous studies (e.g. Boxtel et al., 2002; Roth and Roychoudhury, 1993) have found that collaborative concept mapping has positive cognitive and affective impacts on student achievement in science. For example, Boxtel et al. (2002) study the effect of collaborative CM tasks in learning electricity concepts with 15 and 16-year-old students working in pairs to construct concept maps with a given set of concepts. This concept mapping task served as their introduction to the new course in electricity. Analysis of students' discourse during concept mapping and their concept maps suggests that collaborative concept mapping can provoke productive interaction between students and support meaningful discourse. Moreover, computer-based concept mapping tools offer extensive support for collaborative work during map construction. Such tools facilitate cooperation in the process of concept mapping, allowing partners to view changes made and comment on each other's maps from virtually anywhere (Novak and Cañas, 2006).

Concept Mapping as an Assessment Technique

Even during its "birth" in the early 1970s, the use of concept map as an assessment technique for tracing students' conceptual development was implied (Novak, 2005). Since then, many efforts have been made to explore the concept map's use in diagnosing conceptual understanding and detecting conceptual development. It is generally recognised that concept maps are an effective tool for tracking students' learning through the structural complexity and quality of propositions (e.g. Afamasaga-Fuata'I, 2006; Hasemann and Mansfield, 1995; Pearsall et al., 1997).

Conceptual understanding emphasises relationships, and concept maps are graphic descriptions of these relationships. As such, the concept map is appealing as a tool for evaluating students' conceptual understanding. Students' concept maps depict the organisation of concepts in their knowledge structures. As noted, the proposition is the basic unit of meaning in a concept map, the smallest unit that can be used to judge the validity of a relationship drawn between two concepts (Dochy, 1994; Ruiz-Primo, 2004). Students' maps directly represent their conceptual understanding of a given domain. Misconceptions, usually indicated by an incorrect linking phrase between two concepts, are easy to spot. Similarly, omission of a key concept can suggest a lack of conceptual understanding, lack of conceptual familiarity, or inability to depict a relationship between that concept and other concepts. Ruiz-Primo and her colleagues (Ruiz-Primo, 2004; Ruiz-Primo, Schultz, et al., 2001; Ruiz-Primo and Shavelson, 1996) have worked at length with concept-map-based assessment and offer insights on key issues including assessment tasks, student response formats, response scoring, and validity and reliability problems. I will return to these in detail later.

Concept Mapping Task Formats

As we might expect, concept map assessment tasks appear in various formats in the literature. Novak et al. (1983) consider tasks that include a given list of concepts, that are based on statements in a text, and that include neither given concepts nor given statements. Ruiz-Primo, Shavelson, et al. (2001) provide a more systematic description of mapping formats. They characterised tasks along a continuum from high-directed to low-directed according to who chooses the concepts, who links the concepts, who generates the linking phrases, and who structures the concept map (see Figure 3.2).

Map Components	Degree of Directions	
	High ◄————————————————► Low	
Concepts	Provided by assessor	Provided by student
Lines	Provided by assessor	Provided by student
Linking words	Provided by assessor	Provided by student
Map structure	Provided by assessor	Provided by student

Figure 3.2 Degree of directedness in concept-map tasks.

Source: Adapted from Ruiz-Primo, Shavelson, et al., 2001, p. 101

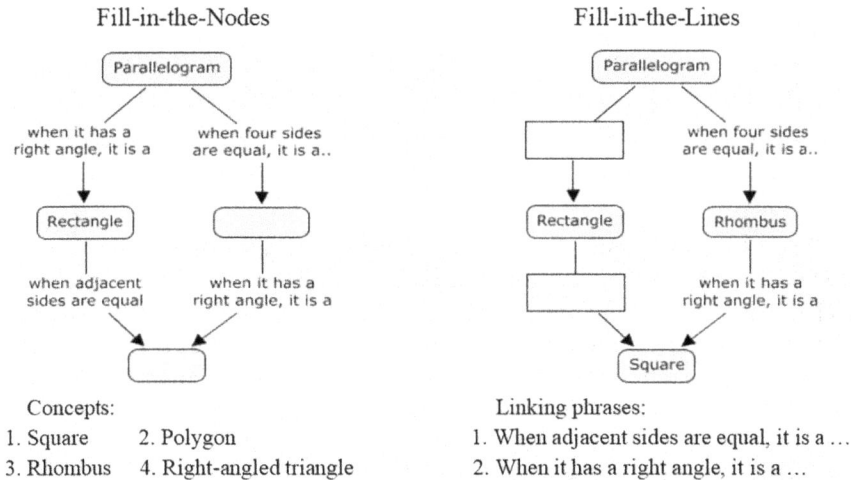

Fill-in-the-Nodes

Concepts:
1. Square   2. Polygon
3. Rhombus   4. Right-angled triangle

Fill-in-the-Lines

Linking phrases:
1. When adjacent sides are equal, it is a …
2. When it has a right angle, it is a …

Figure 3.3 Examples of high-directed concept mapping task: fill-in-the-nodes and fill-in-the-links.

Source: Jin and Wong, 2011, p. 71

If the characteristics of the task fall on the left side of the two-way arrow (high-directed), students' responses will probably be determined more by the task, and it might be easier for teachers to develop a scoring rubric. If the task falls on the right side, students are freer to select concepts, build relationships, add linking phrases, and even decide map structure; this is also known as free-style mapping.

Figure 3.3 presents examples of two high-directed CM tasks: fill-in-the-nodes and fill-in-the-links (Jin and Wong, 2011). Students select correct responses from a list to complete concept maps with missing information. This format is easy to administer, and also easy to grade, such as by counting the number of correctly filled-in nodes or links.

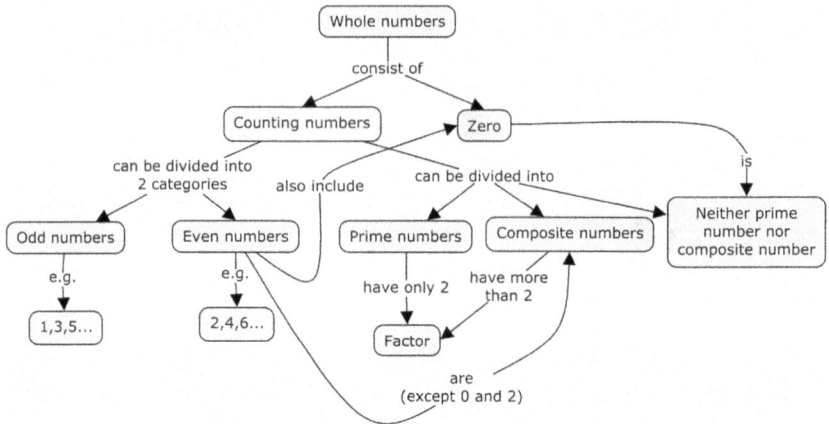

Figure 3.4 An example of low-directed concept mapping task about whole numbers.

Figure 3.4 is the result of a low-directed concept mapping task on whole numbers. In the task, the student identified eight concepts related to whole numbers, included examples, and arranged them in a hierarchical manner. Other students might include different concepts, e.g. real numbers and rational numbers, and organise them in different ways. Considering the diversity of concepts, links, linking words, and map structures, each student is likely to construct a distinctive concept map when presented with such a task.

Studies (e.g. Ruiz-Primo, Shavelson, et al., 2001) have found that the lower the directedness of a concept map-based task, the more opportunities the task provides for revealing students' conceptual understanding. Wallace and Mintzes (1990) and Liu and Hinchey (1996) observed that free-style mapping is an effective vehicle for tracing changes in a student's conceptual development. When no concepts or words are given in the concept mapping task, students are freer to express their own ideas. However, researchers have found that the openness of free-style CM tasks can be undesirable in practice. In comparing a mapping task that included a list of concepts with a task where no list is provided, Ruiz-Primo, Schultz, et al. (2001) found that under the student-selected condition, some students provided related but not relevant or essential concepts, which led to artificially high scores. Similarly, Jin (2007) found that it might be difficult for researchers or raters to develop a reliable scoring system for free style mapping tasks because of the variety of pairs of concepts and relationships that students generate.

Based on these considerations, in the studies reported on in this book, low-directed CM tasks with given concept lists were assigned to participating students. Lines, linking phrases, and map structures were left to students to select.

Methods of Interpreting Concept Maps

In concept mapping studies, once a concept map is completed, evaluating its quality becomes a main concern, especially when the mapping is assigned in the assessment process. Concept maps can be assessed holistically and qualitative based on experts or examiners' judgments or scored along specific criteria.

Scoring is a direct approach to assessing the quality of concept maps. Initially, Novak and Gowin (1984) considered four aspects in scoring: validity of propositions, hierarchy, cross links, and examples, with one point awarded for each valid meaningful proposition, five points for each valid level of hierarchy, ten points for each cross link that is both valid and significant, two points for a cross link that is valid but does not illustrate a synthesis between propositions, and one point for a valid example of specific events or objects. As noted, hierarchy is not always an essential consideration for scoring, and the non-hierarchical spider map or network format can be effective for assessment.

McClure et al. (1999) apply the structural method described earlier along with two other methods: relational scoring and holistic scoring, both of which focus on propositions and concept levels rather than on the structure of students' conceptual knowledge. In the relational method, raters evaluate each proposition in the map and assign it a score from zero to three according to a scoring rubric (see Figure 3.5).

As illustrated in Figure 3.5, the relationship between the connected concepts of the proposition, whether the relationship exists, and whether the linking phrase is compatible with the relationship are considered. The final score of the map is determined by adding the proposition scores. With the holistic methods, examiners rate the overall quality of individual concept maps on a ten-point scale. Each of the three scoring methods allow examiners to compare student-constructed maps with expert constructed maps. In the McClure et al. (1999) study, the expert map was constructed by the professor instructing the students on the mapping topic, and it was provided to raters as a scoring guide. Altogether they applied six scoring methods: holistic, relational, structural, holistic with expert map, relational with expert map, and structural with expert map. Their inter-rater reliability testing indicates that the 'relational with expert map' method was the most reliable of the six scoring methods.

Ruiz-Primo, Shavelson et al. (2001) explored different scoring methods and introduced three types of scores: a proposition accuracy score, a convergence score, and a salience score. The proposition accuracy score is similar to the relational method (McClure et al., 1999) in that the final score is the sum of individual proposition scores, but a different scale is applied. To arrive at a proposition accuracy score, each proposition is scored on a five-point scale, from zero for inaccurate or incorrect propositions to four for excellent or outstanding propositions that reveal deep understanding of relationships between two concepts. The convergence score is the proportion of accurate propositions in a student's map out of all possible propositions on a criterion map. The salience score is the proportion of

```
                    ┌─────────────────────────────┐
                    │  Propositions to be scored  │
                    └─────────────────────────────┘
                                 │
                                 ▼
        ┌─────────────────────────────────┐              ┌─────────────┐
        │  Is there any relationship between │──(No)──────▶│  Assign a   │
        │  the concepts of the proposition? │              │  value of 0 │
        └─────────────────────────────────┘              └─────────────┘
                              │
                           (yes)
                              ▼
        ┌─────────────────────────────────┐              ┌─────────────┐
        │   Does the label indicate a possible│──(No)────▶│  Assign a   │
        │ relationship between the concepts  │            │  value of 1 │
        │       of the proposition?          │            └─────────────┘
        └─────────────────────────────────┘
                              │
                           (yes)
                              ▼
   ┌──────────────────────────────────────────┐         ┌─────────────┐
   │ Does the direction of the arrow indicate a │──(No)──▶│  Assign a   │
   │    hierarchical, causal, or sequential     │         │  value of 2 │
   │  relationship between the concepts of the  │         └─────────────┘
   │ proposition that is compatible with the label?│
   └──────────────────────────────────────────┘
                              │
                           (yes)
                              ▼
            ┌──────────────────────────┐
            │   Assign a value of 3    │
            └──────────────────────────┘
```

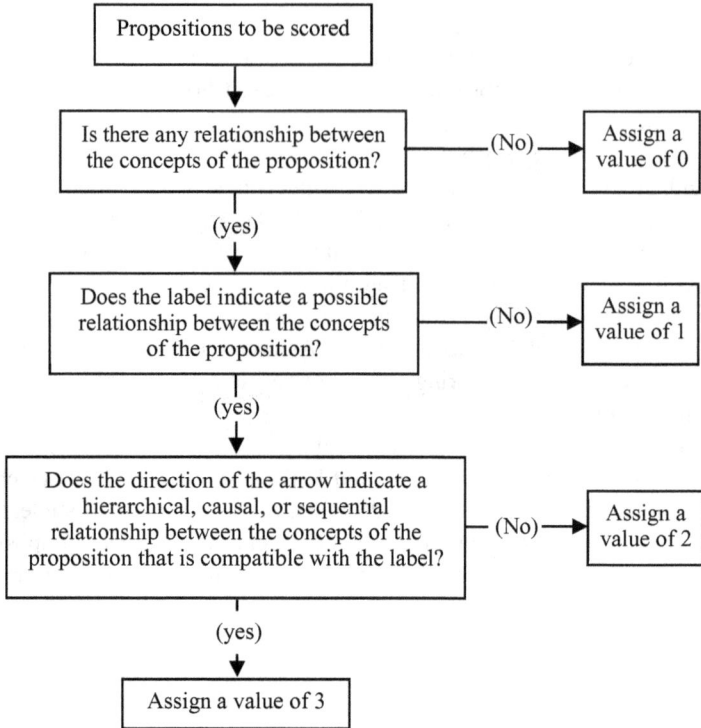

Figure 3.5 Protocol for the Relational Scoring Method.

Source: Adapted from McClure et al., 1999, p. 482

correct propositions out of all propositions on a student's map. Ruiz-Primo et al. found that the proposition accuracy score and convergence score are highly correlated ($r = .95$) and that the correlations between the proposition accuracy score and the convergence score with salience score are also high but lower ($r = .73$ and $r = .75$, respectively). They conclude that the convergence score is the most efficient in consideration of scoring time and reliability. In her study on free-style CM tasks, Jin (2007) considers two additional scores for evaluating the holistic properties of a concept map (similar to Ruiz-Primo et al.'s convergence and salience scores): the proportion of meaningful nodes out of all nodes in a student's map and the ratio of accurate propositions to all meaningful nodes in the student's map. In addition to these scores, Jin focuses on the incoming and outgoing links of individual nodes in a map. A node's incoming links represent the opportunity to activate it from other nodes on the concept map; outgoing links represent the node's power to connect to other nodes. Thus, the sum of the incoming links and outgoing links for a single node may reflect a student's familiarity with the node;

higher scores indicate greater familiarity with the node and that the node is considered more important in the student's cognitive structure.

Different scoring methods emphasise different aspects of concept maps. The same concept map may be scored differently under different criteria. Hence, the corresponding reliability may be affected. The reliability issues of the scoring methods are reviewed in the following section, along with the validity of the concept maps as an assessment tool.

Reliability and Validity Issues With Concept Map-Based Assessment

The reliability and validity of concept mapping-based assessment techniques vary according to the formats of the CM tasks administered and the scoring methods applied.

Reliability

Reliability is defined as 'the extent to which students' assessment results are the same when (a) they complete the same task(s) on two or more different occasions, (b) two or more teachers mark their performance equally on the same task(s), or (c) they complete two or more different but equivalent tasks on the same or different occasions' (Nitko and Brookhart, 2007, p. 67). Most studies that consider the reliability of concept map-based assessments deal primarily with inter-rater reliability and agreement (e.g. Herl et al., 1999; McClure et al., 1999; Ruiz-Primo, Schultz, et al., 2001) and part (b). Only a few studies have conducted equivalency tests with different concept mapping formats (Ruiz-Primo, Schultz, et al., 2001; Yin and Shavelson, 2008), the equivalency of different scoring methods (McClure et al., 1999), and checked for stability across test occasions (Lay-Dopyera and Beyerbach, 1983; Yin and Shavelson, 2008).

Ruiz-Primo, Schultz, et al. (2001) examine the inter-rater reliability and equivalency of two types of CM tasks: fill-in-the-map and construct-a-map-from-scratch. As the former is a high-directed mapping task, scoring it is straightforward, and it tends to achieve high reliability. However, the high reliability is attained at the cost of validity (Yin et al., 2005). The fill-in-the-map task imposes a structure on the relationships between the concepts. Its capacity to reveal students' conceptual understanding and knowledge structure is limited. Construct-a-map-from-scratch is a low-directed task that may better reflect the difference between the students' understanding of the concepts but it is much harder to score. Concerning the reliability of scoring, Ruiz-Primo et al. found that the proposition accuracy score, convergence score, and salience score all had inter-rater reliability coefficients over .90. Ruiz-Primo et al. also correlated students' scores on the fill-in-the-map and construct-a-map tasks and used the convergence scores of the construct-a-map task in the correlation analysis. It turned out that the correlation coefficient between the fill-in-the-map scores and the construct-a-map scores was low

($r = .48$, averaged across types of scores), suggesting that the two mapping techniques may address different aspects of students' conceptual understanding. Yin et al. (2005) conducted inter-rater reliability testing on the other two types of construct-a-map tasks: construct-a-map with created linking phrases (task C) and construct-a-map with selected linking phrases (task S) that were scored using the proposition-accuracy method. They calculated inter-rater reliability at .92 for task S and .81 for task C, which is consistent with the assumption that task C is more open-ended and more challenging to score than task S. Task C and task S were found to be not equivalent in assessing students' conceptual understanding.

In investigating the use of concept mapping for individual assessment with undergraduate students from two sections of an introductory education course, one of Lay-Dopyera and Beyerbach's (1983) goals was determining whether students' knowledge and understanding of a topic, as assessed on a concept map, remained constant across time. The CM tasks were free-style, with only the focus concept given. Half of the participating students were randomly assigned to a concept mapping task on *teaching* and the other half to a concept mapping task on *classroom management*. Students then constructed concept maps again on the same topic after a three-week session on *classroom management*. Their concept maps were analysed in terms of the number of items, number of levels, and number of item streams (branches) they contained. The results of the quantitative tests (paired-sampled *t* test and correlational analysis) and qualitative analysis suggest that the concept mapping task may not be a highly reliable assessment technique. Although moderate consistency was found in the number of items and the number of branches of the concept maps on *teaching*, the qualitative analysis found considerable differences in the pre-and post-maps. By contrast, the concept maps on *classroom management* presented greater content stability, perhaps because it was the focus of the instruction during that period of time.

Yin and Shavelson's (2008) examination of the stability of eighth-graders mapping performance indicates that stability across test occasions may differ according to the concept mapping task format. They first administered a construct-a-map with created linking phrases task (task C) and then a construct-a-map with selected linking phrases task (task S) after seven weeks. No instructional intervention related to content assessed was involved. Four sequences of the tasks were examined: from C to S, from S to C, from C to C, and from S to S, with four groups of students. The discussion focused on CC and SS sequences. Three facets were considered: *student, proposition*, and *occasion*. Yin and Shavelson's (2008) found that the variance component patterns in CC and SS were similar. The interaction of the three facets comprised the greatest proportion of variance: 62.4 percent for task C and 50.3 percent for task S, followed by the interaction of *student* and *proposition*: 16.2 percent for C and 25.3 percent for task S, followed by *student*: 10.4 percent for task C and 18.6 percent for task S; followed by *proposition*: 9.1 percent for task C and 5.1 percent for task S. Their analyses suggest that the openness and flexibility of task C might cause more randomness from one occasion to another than task S. Task S task produced a more consistent measure of students' performance than task C.

Validity

Validity is defined as the 'appropriateness of the interpretation and use made of the results of an assessment procedure for a given group of individuals' (Linn and Miller, 2005, p. 70). In the case of concept-map-based assessments, the focus is on the degree to which the concept map scores reflect students' conceptual understanding.

Correlations between concept map scores and the scores of other measurements of student achievement have been studied at length. Researchers have hypothesised that concept map scores would not correlate well with scores from other traditional tests because they more directly measure students' knowledge structure or the connectedness of their understanding, which is not tapped into with traditional tests (Novak and Gowin, 1984; Shavelson et al., 2005), and some findings supported the hypotheses. Novak et al. (1983) tested correlations between seventh- and eighth-graders' concept map scores and their scores on six school achievement measures (SAT Reading, SAT Math, SCAT Verbal, SCAT quantitative, final examination grade, and final course average). They constructed concept maps using a given list of concepts. The maps were scored on *propositions, hierarchy, branching,* and *cross-links.* Correlations between map scores and the other scores were close to zero, suggesting that concept maps measure different types of achievement than those school tests measure. In another case, Williams (1994) compared the results of university students' free-style concept mapping task with an example-nonexample task. Both tasks are related to the concept of *function.* Correlation (in terms of percentage) between individual student's concept lists with the experts' list and students' total scores on the example-nonexample task were calculated. The correlational analysis indicates that the aspects of conceptual understanding measured by the two tasks has no relationship. Qualitative analysis of the two tasks, however, indicates that concept maps can reveal information on students' misunderstandings which is not readily obtained from the example-nonexample task.

Other studies do report higher correlations between concept map scores and traditional test scores. In Rice et al.'s (1998) study, Grade 7 science students drew concept maps at the end of each unit that were to include 20 to 30 expert-selected concepts. The scoring of the concept maps focused on the *propositions,* excluding *hierarchy* and *branching* (as did Novak and Gowin, 1984). The correlations between the concept map scores and the corresponding multiple-choice test scores ranged from .41–.70, with an average of .55. The researchers then argued that the high correlations provide strong evidence of the validity of the concept map scores. In their study with high school chemistry students, Ruiz-Primo, Schultz, et al. (2001) found correlations of .37, .65, and .44, respectively, between fill-in-the-nodes, fill-in-the-lines, and construct-a-map-from-scratch mapping task scores with multiple-choice test scores. On one hand, this indicates that validity differs among mapping techniques. On the other hand, the moderate correlations suggest that the aspects of students' achievement measured by the CM tasks and the multiple-choice tests are somewhat interrelated, evidencing the content validity of the mapping tasks.

In a study with freshman biology students, Plummer (2008) investigated the criterion validity of two free-style construct-a-map tasks with given lists of concepts by considering the Pearson correlations between scores on their concept map and the scores on their essays and interview transcripts, which are known valid measures. In the first task, participants were directed not to add extra concepts and instructed that they would be penalised with negative points for incorrect propositions. In the second task they were permitted to add related concepts to the given ones and informed that no negative points were assigned for inaccurate propositions. Accuracy, completeness, and relevance of the propositions were considered in scoring. Both the mapping tasks correlated highly with the essay and interview scores (r ranged from .62 to .81).

These findings on the validity of CM tasks may be inconsistent not only due to different formats and the scoring methods but also because participants in those studies were of different grade levels and the CM tasks were on different subjects or topics. Validity issues in the use of concept mapping as an assessment technique should be approached with caution.

In the light of the inconsistent conclusions on the validity of concept mapping as a tool for assessment of students' conceptual understanding, in the present study, I compare the students' concept maps with their responses in definition-example-nonexample tasks and traditional paper-and-pencil tasks (see Chapter 6), thereby contributing to the conversation on the concurrent validity of concept mapping as an assessment technique.

Concept Mapping as a Technique for Assessing Conceptual Understanding in Mathematics

As noted, although Novak (2006) has persuasively argued that the use of concept mapping in science education has important implications for mathematics education since the disciplines share the same psychological and epistemological factors, concept mapping has not been used extensively in teaching, learning, or assessment in school mathematics. Nonetheless, a number of studies addressing the application of concept mapping in mathematics have reported positive results toward its potential integration in mathematics education. This section calls attention to the relevant studies on the use of concept mapping as a technique for assessing conceptual understanding in mathematics, including studies on training and preparation for concept mapping, types of CM tasks, data collection, and key findings.

Mansfield and her colleagues conducted multiple studies on concept mapping in mathematics. In their 1995 paper, Hasemann and Mansfield reported on two such studies, one (Study I) with Grade 4 and Grade 6 primary school students and the other (Study II) with Grade 8 secondary school students. Participating students were not trained in concept mapping prior to embarking on the mapping task. They were instructed step-by-step during the mapping, i.e. directed to position the concepts on the given sheet of paper, position closely related concepts together, and label relationships between concepts. As the sample student maps presented in the paper illustrate, the maps constructed by the fourth-graders did not meet Novak's definition of a concept map; for example, closely related concepts were placed within a single circle and some students did not include links between the concepts. In the study reported in Mansfield and Happs (1991), students were trained in constructing a concept map. After the training, they worked in pairs with a given list of 14 concepts on two-dimensional shapes to jointly construct concept maps. Students were generally able to organise the concepts into the basic concept map form, although the links in the maps were not arrowed and some were not labelled. Notably, these primary school students were even attentive to the hierarchies of the given concepts.

Hasemann and Mansfield's (1995) Study II and the studies by Mansfield and Happs (1989a, 1989b) reported on different stages of the same project with a group of Grade 8 students. In the project, concept mapping was employed to probe students' understanding before and after a teaching programme on parallel lines with a pre- and post-test design. To prepare students for the concept mapping task, the researchers modelled a student-constructed concept map on fractions and presented positive and negative proposition examples. After this, students constructed a map with a given list of ten concepts related to parallel lines. Although concept mapping is often decidedly difficult for younger students, and brief introductions may not sufficiently prepare them to construct informative concept maps, most of the participating students were able to construct a concept map. Moreover, researchers captured meaningful information on their conceptual understanding and conceptual development through their analysis of the concepts and propositions students presented in their maps. These findings encouraged further studies with younger students.

Like Mansfield and Happs (1989a, 1989b and 1991), numerous other researchers have used free-style mapping tasks with lists of given concepts (Bolte, 1999; Lapp et al., 2010; Meel, 2005; Roberts, 1999). While the given lists in Mansfield's mapping tasks included only ten to 14 concepts, these later studies included 20- and 30-concept lists. It may be that greater numbers of given concepts invite greater difficulty in concept mapping. For primary and secondary students, especially when they are first introduced to concept mapping, it is suggested that they start with relatively small numbers of concepts, for example, 14 as in Mansfield's tasks, in the mapping.

Various methods have been used to analyse concept maps constructed with given concepts. Mansfield and Happs (1989b) count the number of correct propositions and incorrect propositions and the number of concepts included. Bolte (1999) applies a holistic scoring criterion focused on how the maps are organised and the accuracy of the propositions. Other researchers discuss students' concept mapping performance in detail, for example, the specific propositions students construct, which of these are incorrect, and which given concepts are isolated or omitted.

Lapp et al. (2010) focus on the strength and richness of the connections students represent in their maps. When fast-forwarding video data, they noticed that the students' appeared to be using discrete thinking skills in constructing their concept map. Students' seemed to be briefly reflecting on the given concepts and then constructing a portion of the concept map 'in a sudden burst' (Lapp et al., 2010, p. 8). Lapp, et al. refer to such portions as *clumps*. They then distinguish between inter-clump links and intra-clump links. Concepts within a clump are considered strongly connected. Adjacency matrices were employed to record the number and lengths of the paths between concepts in a pair. I refer to these methods when analysing the concept maps described and presented in this book.

Free-style CM tasks with no given concepts have also been tested (Afamasaga-Fuata'I, 2006, 2009a, 2009b; Schmittau, 2009; Schmittau and Vagliardo, 2009; Jin, 2007; Hough et al., 2007; Williams, 1994, 1998). However, as mentioned, mapping tasks in this format may be too open to facilitate students' concentration on the intended domains. For example, Jin (2007) reports that some secondary students included the concepts *moon cake* and *rabbit* in their concept maps of *circle*. Moreover, challenges to developing reliable scoring systems for this type of mapping may make its large-scale adoption for assessment purposes impractical.

Afamasaga-Fuata'I (2006, 2009a, 2009b) conducted a series of case studies using concept maps to trace student teachers' conceptual knowledge of matrices and systems, length and volume, and fractions. After a period of study, the student teachers generated lists of concepts for the topic and constructed concept maps illustrating their understanding of concepts' inter-connectedness and presented their maps to the class or researcher. Through discussions and negotiations, the student teachers further revised and expanded the maps. Progressive maps were collected and compared by the researcher. In the Afamasaga-Fuata'I (2006) study, concept maps are analysed systematically on structural complexity, nature of the nodes, and valid propositions. She counts the hierarchical levels, multiple-branching nodes, cross-links, uplinks, sub-branches, and valid propositions present in the maps. Unlike Novak and Gowin (1984), she does not assign collective scores to the concept maps. She focuses more on the qualitative information in the map. In her studies (2009a, 2009b), Afamasaga-Fuata'I concentrates on the propositions in concept maps. Improvements in student teachers' concept maps are documented.

Schmittau (2009) and Schmittau and Vagliardo (2009) use concept map to develop and assess student teachers' conceptual understanding with a focus on the final maps constructed, in contrast to the progressive analysis approach taken by Afamasaga-Fuata'I. Student teachers' final concept maps were complex, often with over 40 nodes and were difficult to interpret as a whole. Hence, the researchers divide the maps into sections and analyse the sections separately. They found that the concept maps the student teachers drafted were reflective of their conceptual understanding of the material and revealed both strengths and deficiencies in that understanding; differences among student teachers' conceptual understanding were also captured (Schmittau and Vagliardo, 2009).

Although positive findings are reported, the CM tasks and data analysis methods presented in those papers are not practical for adoption by school teachers because it is difficult for younger students to construct maps with so many concepts. Moreover, teachers may find analysis of concept maps time-consuming, and they cannot really be expected to apply it at the whole-class level. These case studies do though provide useful information on analysing concept maps from a qualitative perspective.

Rather than asking subjects to construct concept maps, Chinnappan and Lawson (2005) developed concept maps based on video-taped and transcribed interviews with two experienced teachers. The two teachers were interviewed individually and made explicitly aware that the interviews were intended to reveal their understanding of geometric topics and of the teaching of those topics. The researchers categorize the teachers' knowledge of geometry and knowledge of geometry for teaching geometry according to definition features, related features, applications of the knowledge, and other, corresponding with the four schemas presented in the researcher-constructed concept maps. Teachers illustrated connections within and between schemas. The nodes, links, and the structure of the concept maps together presented information about the teachers' understanding of the focus concept visually.

Figure 3.6 is a researcher-constructed concept map from their study. The dotted box on the upper left is a space to record *applications* of the given concepts. In this sample, the box is empty, indicating that the teacher did not discuss applications of the focus concept *square* during the interview. The concept map generally indicates that the teacher had a good understanding of the properties of *square* and the connections between *square* and other geometric figures but may not have been as familiar with its real-world applications. Differences between the two teachers' understanding of the focus concepts were captured through comparison of their concept maps. As mentioned, Novak (2005) reported a similar study using researcher-constructed concept maps to trace students' conceptual development in science based on interviews with students. These studies suggest that even when used indirectly, concept maps have multiple advantages in terms of representing teachers and students' conceptual understanding of certain topics.

The CM tasks in the studies reviewed in this section are geometry-related, e.g. *parallel lines*, *length*, and *volume*, or algebra-related, e.g. *linear algebra* and *functions*. Although Bartels (1995) and Bolte (1999) study concept mapping for both geometric and algebraic topics, they do not investigate the differences between the use of concept maps for the two topic domains. The meta-analysis by Nesbit and Adesope (2005) indicates that the effect size of concept mapping as a learning strategy differs by topic. Possible topical differences in the use of concept mapping as an assessment technique may also exist.

It is important to understand and control the effects of concept mapping skills on the quality of concept maps. One way to do so is to provide mappers with extensive training in concept mapping. This aspect has been adequately studied in the general literature in concept mapping. However, few of the mathematics studies have provided much effort on training, not to mention the effectiveness of the training.

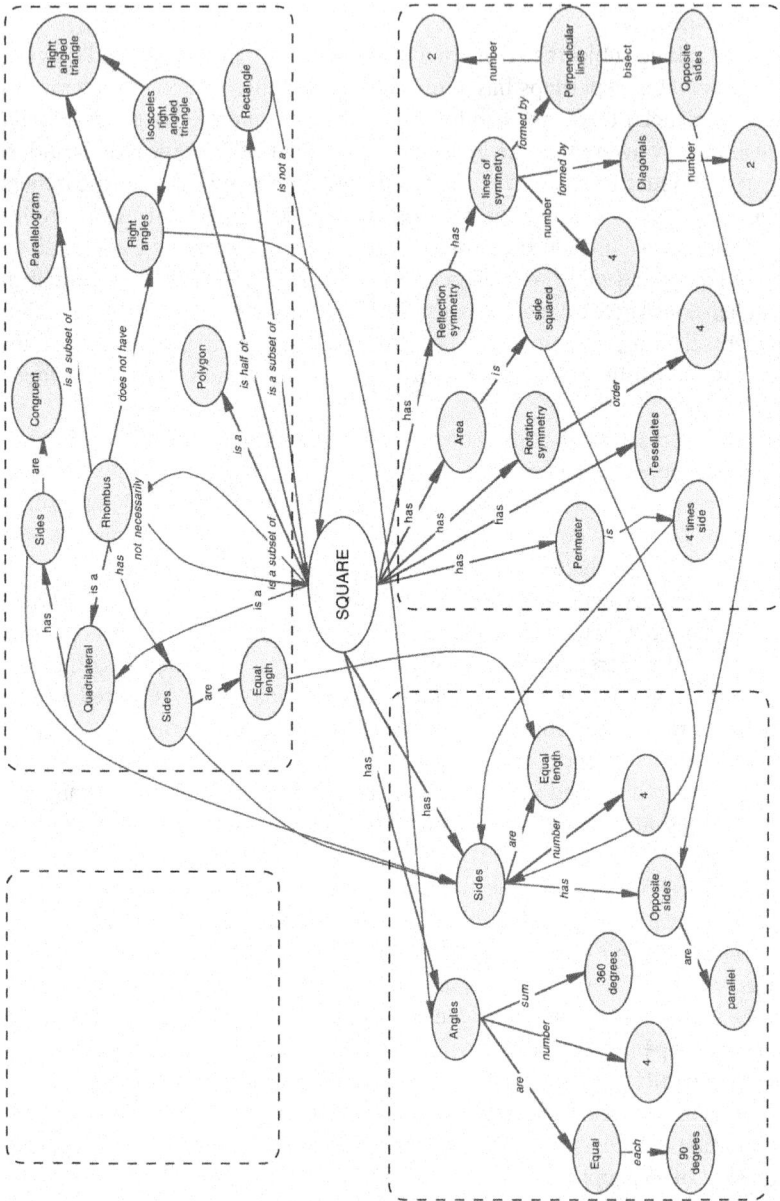

Figure 3.6 A researcher-constructed concept map.

Source: Chinnappan and Lawson, 2005, p. 211

Attitudes Toward Concept Mapping

Students and teachers' attitudes toward concept mapping are a key issue in determining how widely it will be applied in school mathematics education as an instructional method, a learning strategy, and an assessment technique. As studies on students or teachers' attitudes toward concept mapping in mathematics appear unavailable, I present studies on their attitudes toward concept mapping in general, to further guide our exploration.

In Kankkunen (2001), primary students constructed concept maps on topics in biology, history, geography, religion, and environmental education over four years. Kankkunen designed a 15-item questionnaire to address their attitudes toward concept mapping in general after their extensive, prolonged use of concept maps. Students were 9–12-years-old when they completed the questionnaires. The items were developed from notes from the researcher's research diary. Most students expressed primarily positive attitudes toward concept mapping. For example, 96 percent found it helpful to them in their studies and 91 percent agreed that concept mapping helped them organise their learning; also, some 52 percent found concept mapping labour intensive. Such results indicate that concept mapping can be very helpful and also very demanding.

Students' positive attitudes toward concept mapping have also been found after relatively brief exposure to mapping. In his study on concept mapping and achievement in science, Mohamed (1993) surveyed a group of students at a government school in Singapore on their concept mapping attitudes using a researcher-designed questionnaire. The questionnaire includes 29 items which examine three aspects of concept mapping attitudes: preference for concept mapping, usefulness of concept mapping, and self-appraisal and self-evaluation using concept mapping. Participating students were instructed on concept mapping and required to draw concept maps before and after each session. The Attitudes Toward Concept Mapping Questionnaire (ATCMQ) was administered after two 3.5-hour training sessions. Most students reported a positive attitude. Particularly, a significant correlation value of 0.56 ($p = .005$) was presented between their post-test results and their attitudes as measured by the questionnaire. This may indicate that students who achieve better post-test scores may also hold more positive attitudes toward concept mapping.

In Wang (2005), chemistry majors enrolled in a general physics course at a normal university in Zhejiang, China were introduced to concept map and were required to draw a concept map after completing each chapter. In between, teacher-constructed concept maps were modelled and the connections and organisation of the model maps were explained. At the end of the semester, open-ended questions were assigned to collect the students' ideas about the concept mapping, and a questionnaire was designed based on students' responses. While students generally found concept mapping to be a useful learning strategy and an effective instructional tool, 70 percent of them hoped that their teacher would not ask them to construct concept maps the next semester. They would rather be provided with well-made expert-constructed maps. Most students acknowledged that they knew something about how to construct a concept map but also that such knowledge was not at all extensive. Moreover, most students did not consider concept map a helpful tool for problem solving, which is considered the most important issue in physics. They reported that teachers should spend more time teaching problem-solving. This reaction is understandable because concept mapping does not usually involve operations or problem solving.

In their report on a study on the feasibility of concept maps for measuring content achievement in reading expository texts, Anderson and Huang (1989) reported that their eight-graders participants expressed negative attitudes toward concept map, particularly as an assessment tool in schools. Participating students learned mapping techniques requiring them to convert their ideas into propositions and arrange them into concept maps. Results from the attitude questionnaire administered at the end of the mapping sessions indicate that despite substantive diagnostic value in terms of what students do and do not understand, students did not, as noted, hold positive attitudes toward the concept mapping tests they took in school.

These results indicate that students' attitudes toward concept mapping may vary according to the subject, topic, or purpose (e.g. as a learning strategy or as an assessment technique). This study conducted by the present author as described in Chapter 8 of this book examines students' attitudes toward concept mapping as a technique for assessing conceptual understanding in mathematics, together with mathematics teachers' attitudes toward concept mapping, which have not previously been investigated extensively.

Summary

In this chapter, relevant issues concerning the use of concept mapping as an assessment technique, students' attitudes toward the concept map, and studies related to both conceptual understanding and the concept map in mathematics have been reviewed. This review indicates that further study is need in the following aspects: (1) concept mapping training programmes, (2) concept map interpretation and analysis, (3) the validity and reliability issues with the concept map as a tool for assessing conceptual understanding, and (4) questionnaires for measuring attitudes toward concept mapping.

4 Developing a Training Programme for Concept Mapping in Mathematics

Training on Concept Mapping in the Literature

In many countries and regions, concept maps have not been used extensively in schools. Especially in such cases, researchers should introduce concept map to students and teach them how to construct one prior to implementing concept mapping for teaching, learning, or assessment purposes.

Novak's hierarchical concept map procedure (Novak and Gowin, 1984; Novak, 1998) is widely used. However, the hierarchical structure presents some limitations. For example, it increases students' cognitive load during concept mapping. Accordingly, the procedure has been adjusted and adapted for more general use (Ruiz-Primo, Schultz, et al., 2001; Wallace and Mintzes, 1990).

That being said, constructing a concept map generally involves the following five major steps:

1 *Focus question*: Formulate a focus question that addresses the problem, issues, or knowledge domain.
2 *Key-concept list*: Guided by the focus question, list 10 to 20 pertinent concepts.
3 *Connection*: Arrange the concepts and connect them with arrows. Continue this process until all related concepts are linked.
4 *Linking phrases*: Label the arrows with one or more linking phrases to define relationships between connected concepts.
5 *Reorganisation*: Rework the structure of the map, which may include adding, deleting, or changing the location of concepts to increase the legibility of the final work.

The focus question is usually assigned by researchers or teachers. Adjustments can be made to the other steps according to subject domain, mapping format, student grade level, and the purpose of mapping task. For example, in less challenging high-directed tasks, such as fill-in-the-nodes where students need only supply linking phrases to explicate the given connections, Steps 2, 3, and 5 may not come into play.

DOI: 10.4324/9781003269373-4

Concerning concept mapping training, researchers have expended various levels of effort with diverse training methods (e.g. Edwards and Fraser, 1983; Jin, 2007; Ruiz-Primo, Schultz, et al., 2001; Wallace and Mintzes, 1990; Williams, 1998). Some suggest that students are capable of constructing valid concept maps in little time with limited practice. For example, in Edwards and Fraser (1983) and Williams' (1998) work with high school and college students, respectively, researchers began with a brief training on constructing concept maps and then proceeded directly to the mapping task. In Jin's (2007) study on students' mapping on function conducted in China, she found that having been provided with only one example, the participating secondary school students were able to construct information-rich concept maps.

Other researchers have had limited success with shorter training sessions. Wang and Dwyer (2006), for example, found that the 50-minute workshop they conducted on concept mapping with participating undergraduate students did not develop their mapping skills sufficiently to prepare them for actual mapping in the main study. Accordingly, the researchers advocated more vigorous training.

The training method developed by Wallace and Mintzes (1990) is quite elaborate. They tested the concurrent validity of concept map as a tool for exploring changes in university students' conceptual understanding in biology. They conducted six 75-minute sessions, one for each of the following: training, practice, review, pre-testing, instruction, and post-testing. The first three sessions were intended to develop students' concept mapping skills. In the training session, operational definitions of concept map—concepts, propositions, principles, theories, relationships, hierarchy, cross-links, and general-to-specific examples—are introduced, followed by sample student maps, scoring criteria, and sample scored maps. The subsequent practice session, wherein students practiced the mapping technique, includes a card-sorting exercise for generating concept maps (Novak and Gowin, 1984), a review and discussion of the student-generated maps, and a homework assignment to develop and submit a concept map based on a chapter in a middle school science textbook. In the review session, the maps submitted by the students in the last session were scored by the researchers and the scored maps were returned to the students. The scoring method was re-examined, and problematic areas were then discussed. The students spent nearly four hours studying concept mapping before the actual practice. Revealing the scoring criteria and other discussions can also help students understand the requirements of mapping tasks. However, Wallace and Mintzes did not report on the results or effectiveness of the training.

Ruiz-Primo et al. (2001) formally evaluated the effectiveness of a 50-minute training session applied in their study exploring concept mapping and high school students' conceptual understanding of chemistry. In the initial section, definitions, components, and applications of concept maps were presented to students along with examples of hierarchical and non-hierarchical maps. Subsequently, students learned about identifying relationship between concepts in a pair, constructing propositions, evaluating maps for quality, redrawing maps, and constructing a map collaboratively; next, they practiced mapping with a given list of concepts (in this case nine). The training ended with a class discussion on students' questions about mapping after having practiced individual and collaborative mapping. At the end of the training, researchers evaluated the effectiveness and concluded that the training programme was successful in training students to construct concept maps. After the training session, students worked alone and each constructed a map using two given lists of concepts on chemistry topics. To evaluate the effectiveness of the training, researchers randomly selected 25 percent of the student-constructed maps and evaluated them for use of the concepts in the list, use of labelled links, and accuracy of propositions. Over 94 percent of the students had used all of the concepts in the list, all students had included labelled lines, and over 96 percent had included one or more valid propositions. Ruiz-Primo et al. concluded that the training effectively prepared the participating students to construct concept maps. I further trialled Ruiz-Primo et al.'s evaluation method in the preliminary study reported in the following section; however, the results of that study suggest that such training is not sufficient to differentiate among students' mapping abilities.

The literature does not adequately address the issues related to training on concept mapping, especially how much and what types of training is sufficient and for which levels of students. Understanding how students acquire and develop their mapping skills is a pertinent issue that should be clarified before going deeper into any study using concept map as an assessment technique. Given this gap in the literature, I put much effort into the training sessions and into examining students' mapping skills.

The Preliminary Training Programme

A preliminary study was first conducted to explore the training techniques described earlier in the literature and to develop a more elaborate training programme.

Participants

Ten (A-J) students were recruited by convenience sampling. Students A-H had been invited to participate by their mathematics teacher at a government secondary school in Singapore (because they were articulate), and they accepted the invitation. Students A (male) and B (female) were Grade 7 students, and Students C-H (all males) were Grade 8 students. Some of them had heard about concept map or had seen similar maps (e.g. mind maps), but they all claimed that they did not know how to use or construct concept maps. Student I, a male Grade 7 student, and J, a female Grade 9 student, are the grandchildren of a mathematics professor at the researcher's university, Singapore. Student I had experienced difficulty learning mathematics and knew little about concept map before the study. Student J was a top student in her class and had used concept mapping for her own study since Grade 4 when she was first introduced to the technique by her science teacher.

Training Methods and Students' Mapping Performance

Group 1: Simple Training for Students A and B

For Group 1, training began with a ten-minute session on Novak's definition of concept map, focusing on the three key features: nodes representing concepts, arrowed and labelled links reflecting directional connections, and linking phrases describing the relationships between linked concepts. This was followed by a five-minute concept mapping demonstration with eight given concepts: whole numbers, counting numbers, zero, prime numbers, composite numbers, odd numbers, even numbers, and factors. Students were advised to proceed stepwise together in jointly constructing a concept map, from reading through the list, to positioning each concept on the blank page, to drawing the connections, to adding the linking phrases. They were then each given a blank page and five stickers labelled with the following concept names: isosceles triangle, triangle, acute angle, equal sides, and acute-angled triangle. Each student was advised to proceed stepwise and construct an individual concept map. Their maps met Ruiz-Primo, Schultz, et al.'s (2001) standard indicating effective training. However, most of their linking phrases were very general. When prompted by the researcher, both students added additional links and more detailed information to the linking phrases in their maps. For example, Student B initially constructed the proposition 'isosceles triangle is a kind of triangle'. When asked to explain what he meant by 'a kind of', Student B said that 'isosceles triangle is a triangle with two equal sides; thus, it is a kind of triangle'. The implication is that students knew more than what they included in their concept maps; the brief training had not prepared them to construct concept maps that reflect their full knowledge of the concepts in the given list.

Group 2: Accuracy of Linking Phrases for Students C and D

Given these results, I then conducted a similar training session with Students C and D but placed greater emphasis on linking phrase accuracy and encouraging them to add as many connections and details as they could think of. A 15-minute explanation and demonstration session was presented (similar to the ten-minute session described earlier). They collaboratively constructed a map with the same given triangle concept list described earlier. They then constructed maps individually with a given 12-concept list comprising quadrilateral, angles, parallelogram, square, rectangle, trapezium, rhombus, polygon, symmetry, diagonals, adjacent sides, and opposite sides. Both students completed the mapping task in about 25 minutes. The concept map constructed by Student D (redrawn by the researcher for clarity) is presented in Figure 4.1.

Student D captured the essential aspects of a concept map: connections, linking phrases, and the general hierarchy. However, the links and linking phrases are still overly brief without detailed information, for example, '[*rectangles*] have [*opposite sides*]' and '[*polygon*] has [*angles*]'. As with Group 1, when prompted, the

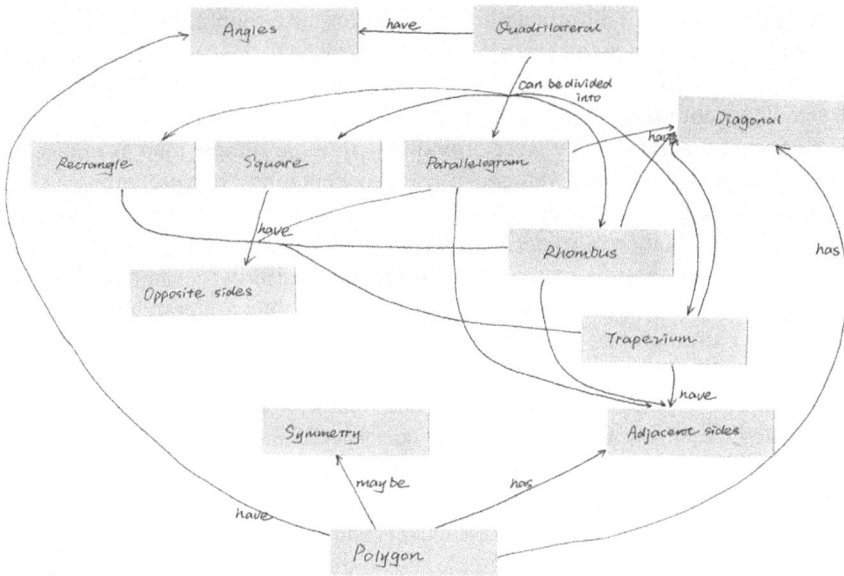

Figure 4.1 Student D's concept map on quadrilaterals (redrawn for clarity).

Source: Jin and Wong, 2010, p. 106

students expanded on the relationships, in this case, the relationships among parallelogram, rectangle, square, rhombus, and trapezium, as in the following exchange:

R (Researcher): Here, you write 'quadrilateral can be split into: rectangle, square, parallelogram, rhombus, and trapezium'. Can you explain a bit more about this proposition?

D (Student D): These five concepts are all quadrilaterals, but they are different from each other.

R: Can you give me an example?

D: Sure. See, this is a trapezium [he drew a trapezium] . . . this is a rhombus [he drew a rhombus beside the trapezium] . . . they are different.

R: So the difference is. . . ?

D: This one [the trapezium] has one pair of parallel lines, but this one [the rhombus] has two.

This conversation reveals that the student's map did not include all of the information that he had in his possession. Perhaps students were treating concept mapping as a drawing task rather than as a test of their conceptual understanding. This suggests that training should include specification of the purpose of the mapping tasks—which may differ by application—and clearer instructions and modelling.

Group 3: Extended Training for Students E, F, G, and H

I planned an extended 50-minute training session for Group 3. In addition to the explanation and demonstration provided for Group 2, this session covered the purpose of CM tasks and included examples of well-constructed and poorly constructed concept maps, along with opportunities for students to practice and discuss concept mapping. They were provided with three sample map examples on the topic *numbers*. Two are well-constructed concept maps: one presenting a clear hierarchy among the concepts with the most inclusive concept at the top level of the map and the other in spider map format with the most inclusive concept in the middle of the map (also see Chapter 3). With these two examples, I aimed to convey that a concept map can be constructed with a hierarchical or non-hierarchical structure, as long as the ideas are clearly expressed and the format is appropriate to the concept. The third sample was a poorly constructed concept map with overly general linking phrases and relatively few mapped connections. I explained why it was considered poorly constructed and asked for students' suggestions on refining it. These examples were intended to encourage students to include as much detail and as many connections as they needed to fully express the extent of their knowledge of the concepts and the connections between them. After a brief discussion on the sample maps, the students performed the same collective mapping task with the given list of triangle concepts. During the mapping process, they reminded each other to focus on the accuracy of their linking phrases, and they adjusted the structure of the created propositions and proposed additional connections.

After the collaborative mapping practice, the students constructed individual concept maps using a given quadrilateral concept list. Student E completed the task in less than 15 minutes, students F and G in around 20, and H in around 40 minutes. I was able to draw more diagnostic information from these maps than from those constructed by Group 2 students. Student F's map is presented in Figure 4.2 and student H's in Figure 4.3.

Both maps are divided into three general layers, with quadrilateral, the most inclusive concept, placed at the top of the map; the five quadrilaterals, parallelogram, rhombus, square, trapezium, and rectangle, are at the intermediate level; and the remaining four concepts describing properties are at the lowest level.

Most of the connections in Student F's map are between the particular quadrilaterals and the property-related concepts, i.e. symmetry, angles, sides, and diagonals. Student F used detailed linking phrases to label the connections. For example, he linked square and adjacent sides with the linking phrase 'has perpendicular' to form the proposition '*square* has perpendicular *adjacent sides*', rather than the less detailed '*square* has *adjacent sides*'. Detailed linking phrases offer more information on students' conceptual understanding and can reveal misconceptions. For example, Student F proposed '*quadrilateral* is not a *polygon*' and '*trapezium* [has] one set of equal *adjacent sides*', suggesting that he was not clear on the relationship between *quadrilateral* and *polygon* and the property of the trapezium. Such information is of value to teachers in better understanding students'

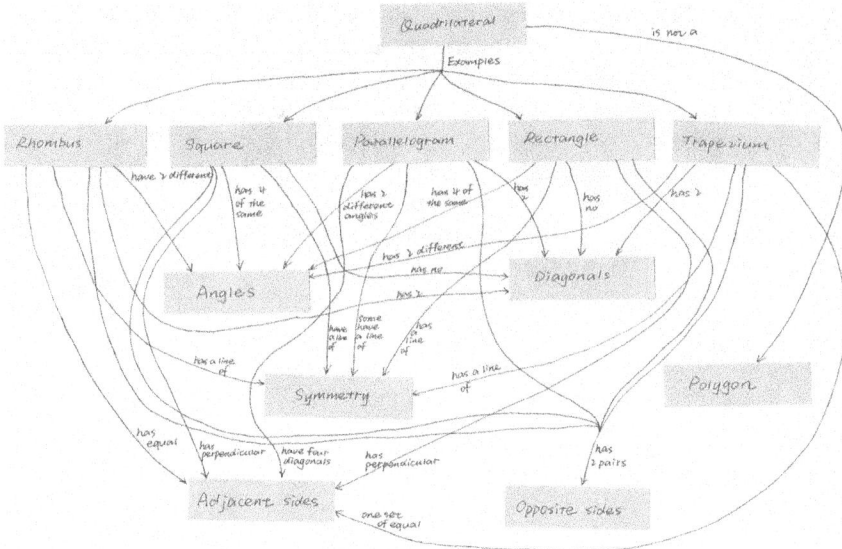

Figure 4.2 Student F's concept map on quadrilaterals (redrawn for clarity).

Source: Jin and Wong, 2010, p. 108

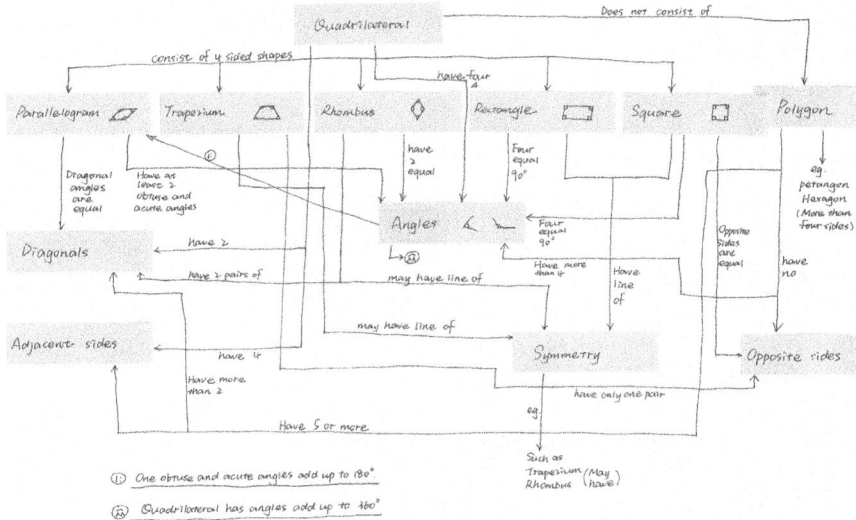

Figure 4.3 Student H's concept map on quadrilaterals (redrawn for clarity).

Source: Jin and Wong, 2010, p. 108

learning difficulties and thereby improving their teaching. Since no connections are illustrated among the concepts parallelogram, rhombus, square, trapezium, and rectangle, no inferences can be drawn about Student F's conceptual understanding of the relationships among the five quadrilateral types.

Student H illustrated many connections between the given concepts. He carefully checked the possible links between each pair of concepts. When he found that there was no more space to neatly add more propositions, he turned to me for help. I suggested he mark the links and write his explanations below the map. His map provides substantial information on his conceptual understanding. Misconceptions are easily detected. For example, the proposition '*trapezium* have only one pair [of] *opposite sides*' in Figure 4.3 indicates that student H may confuse opposite sides with parallel sides, whereas propositions '*polygon* have five or more *adjacent sides*' and '*quadrilateral* does not consist *polygon*' together show that he interpreted polygon as a figure with more than four sides.

Extensive Training for Student I

The researcher worked with Student I individually. He received the same training as Group 3 students. However, in the practice phase, he worked with seven numbers concepts—rational numbers, integers, fractions, number line, absolute value, positive numbers, and negative numbers—which were different from the concepts provided in the training. In the mapping process, he at first copied the structure and linking phrases in the researcher's example, not seeming to recognise that the example and the practice task were on two different sets of concepts. After ten minutes, he had linked only three concepts, two of them incorrectly. His performance indicated that he did not understand the purpose and requirements of concept mapping. The training was not adequate for this student to acquire the mapping ideas.

Brief Training for Student J

Student J has used a form of concept mapping very often for her own study. However, the nodes in her maps were not necessarily concepts, but sometimes phrases or sentences. Moreover, the links were not necessarily labelled. Accordingly, I focused her ten-minute training session on explaining the differences between the method of mapping she had been practicing and the method described in this book, followed by the same stepwise instructions presented to Group 3. After the brief training session, Student J build a concept map with a given list of triangle concepts. During her mapping practice, I asked her two questions: (1) "You wrote '*right-angled triangle* has *acute angle*'. Can you tell me how many acute angles a right-angled triangle has?" (2) 'Any relationship between *isosceles triangle* and *acute-angled triangle*? Can you make a link between them?' The first question reminded her to express the link with thorough and detailed phrases, and the second question reminded her to think deeply about the connections. Prompted by the questions, Student J added more connections to her map and labelled them more accurately.

In the test for effectiveness in expressing conceptual understanding, Student J's given list on quadrilaterals was different from the list given to Group 2 and 3. Her list included nine rather than 12 concepts since Group 3's experience suggests that a single page may not have enough space for students to fully express their knowledge on 12 concepts. Initially, she arranged trapezium, rhombus, parallelogram, kite, square, and rectangle at the same level as Student E and Student F did. After some time, she realised that there was also a hierarchy among these six concepts. She erased all the connections and redrew the map. The final map is presented in Figure 4.4.

As a whole, this map indicates comprehensive understanding of the nine concepts. Student J completed the first draft and this final map in 15 minutes.

This training experience indicates that different training methods have different effects on students' concept mapping. Even with the same training provided, students' mapping skills might differ. These preliminary studies also suggest that 45-minutes might not be sufficient training time for secondary students, as most students did not fully express what they knew about the concepts and connections

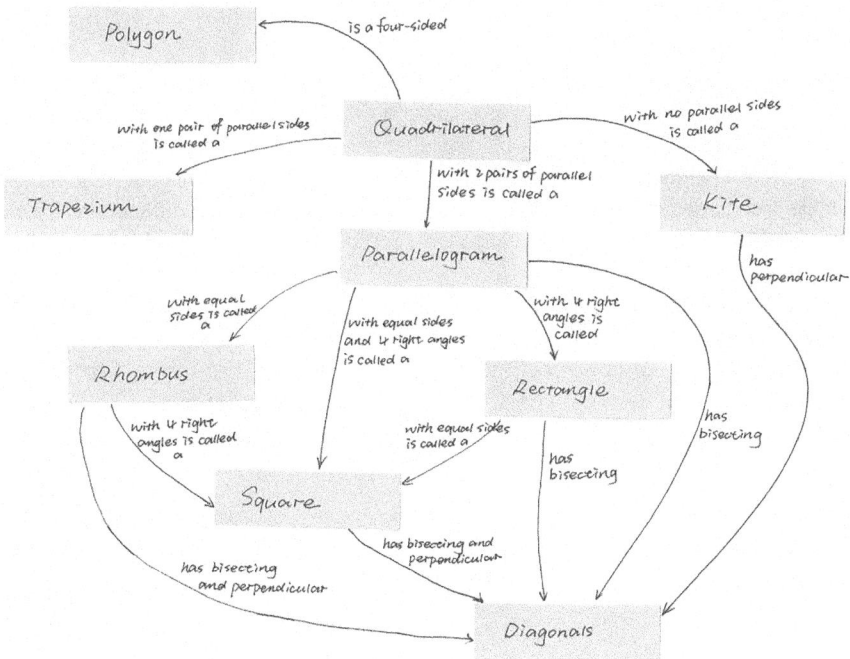

Figure 4.4 Student J's concept map on quadrilateral (redrawn for clarity).

Source: Jin and Wong, 2010, p. 110

through their concept maps. Students need more practice to develop concept mapping competence. Such competence is required before their maps can be considered accurate reflection of their concept understanding.

Defining Concept Mapping Skills

Concept mapping skills (CMS) have a specific meaning in this book, derived from the findings obtained from the preliminary training experiments just described and together with other training considerations raised in the literature, in particular, the work of Ruiz-Primo and Shavelson (1996) and Williams (1998). CMS comprise three key concept mapping skills defined as follows:

Skill 1: *Statement transformation* refers to the ability to transform a given statement into a diagrammatical proposition. It requires keeping the meaning of the proposition consistent with the meaning of the given statement. Students must be able to first identify key concepts from the given statement, then capture their relationships, and then express the relationships through linking phrases. Basic examples are presented in Figure 4.5.

Statement 1 is straightforward. However, Statement 2 can be mapped differently, as illustrated by two examples: one with three nodes and the other with two nodes. As long as the meaning of the given statement has not been changed, it does not matter how nodes (concepts) are used. Students are not expected to express their ideas at this stage.

Skill 2 is *simple free association*. While statement transformation involves the ability to work with given concepts, simple free association involve generating connections and detailed linking phrases themselves. To develop simple free association skills, students should work with two to four given concepts. High-level simple free association skills fully express what they know about the connections between the given concepts. No concept should be left isolated (unconnected to other concepts) unless it has not been taught. Since different students may connect the same concepts differently, free association propositions will expose their understanding of the concepts. Consider the examples presented in Figure 4.6.

Proposition 1 does not show how a quadrilateral is different from a rectangle. Comparatively speaking, Proposition 2 more accurately expresses the connection between the two given concepts. Proposition 3 indicates a misconception.

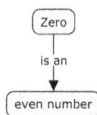

Statement 1: Zero is an even number.   **Statement 2:** A rectangle is a quadrilateral with four right angles.

Key concepts: zero, even number.   **Key concepts:** rectangle, quadrilateral, angles.

Proposition:

Zero

is an

even number

Proposition:

Rectangle

is a   has four right

quadrilateral   angles

Quadrilateral

when it has four right angles, it is a

rectangle

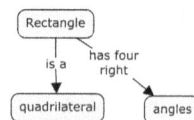

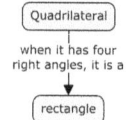

Figure 4.5 Examples for the first CMS statement transformation.

Source: Jin and Wong, 2010, p. 111

Key concepts given: quadrilateral, rectangle.

Proposition: (1)   (2)   (3)   (4)

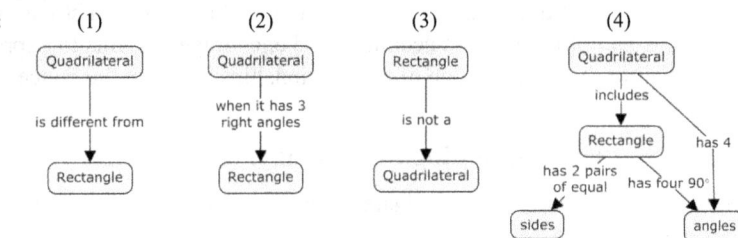

Figure 4.6 Examples for the second CMS simple free association.

Source: Jin and Wong, 2010, p. 111

Proposition 4 reflects comprehensive understanding of the quadrilateral and related concepts. In the sample mapping (Figure 4.6), two additional related concepts are added. This is acceptable as long as the connections between the given concepts are indicated with correct phrases.

Skill 3: *Extended free association* is built on simple free association skills. It involves the ability to map more than four concepts and requires deeper attention to structural elements, i.e. the hierarchy of the concepts and the organisation of the propositions. Students need to be skilled in expressing conceptual hierarchies as such structures are often required to capture essential interrelatedness among mathematical concepts (Sfard, 1991; Skemp, 1987). They must also become skilled in organising the propositions so that their concept maps are clear and legible.

Refining the Training Programme

Having considered the results of the preliminary training experiments and toward developing students' CMS (as defined earlier), I revised the training for concept mapping in mathematics to include the following aspects:

- *Introduction:* Define the concept map and describe its specific attributes, nodes, arrowed links, and linking phrases.
- *Statement transformation:* Use examples to explain the process of converting statements into propositions without changing the meaning of the statements, followed by practice and comments.
- *Simple free association:* Use examples to model construction of propositions with two to four given concepts, followed by practice and comments.
- *Extended free association:* Use examples to model concept mapping with more than four given concepts. Encourage but do not require students to construct hierarchical concept maps. Ask them to consider the overall

organisation of the entire map and redraw it if necessary. Practice and comments should follow.

- *Discussion:* Summarise common problems that appear in students' practice, focus on linking phrases accuracy, and encourage students to express the full extent of their knowledge of the relationships between the concepts.

Implementing the Training Programme: A Pilot Study

The pilot study aimed to investigate the effectiveness of the training programme for secondary students.

Participants

Thirty-six Grade 8 students, aged 14, in a junior middle school in Nanjing, China, participated in the pilot study. The students indicated that they had no previous knowledge of concept map and had not participated in similar mapping tasks.

Training

The class received three 40-minute training sessions on concept mapping for mathematics over two weeks. I followed the training programme outlined earlier. Session One introduced concept mapping and statement transformation. Practice and comments were provided in class. At the end of this session, I assigned statement transformation exercises as homework. The homework was collected one day before Session Two, and we discussed errors at the beginning of Session Two, followed by simple and extended free association skills building. Given the level of complexity involved in building free association skills, the first half of the third session featured a continuation of that training, followed by discussion and summary of key ideas about concept mapping.

Effectiveness of the Training

To test the effectiveness of the training in the pilot study, I designed a Concept Mapping Skills test (CMS-test) (see Appendix A). It comprises three types of tasks designed to measure their statement transformation, simple free association, and extended free association skills. Concepts and linking statements from the students' mathematics textbooks were used. Different concepts than those presented during training to gauge their ability to apply their mapping skills in multiple mathematical contexts were given.

The CMS-test was administered two days after the last training session, and students had 45 minutes to complete it. I assessed their mapping skills using a checklist with the following categories: concepts, links, linking phrases of propositions, and organisation of propositions. For concepts, I checked whether the concepts were correctly extracted from the given statements and whether all the given concepts were included. Students were permitted to add related concepts, but this was not a factor in the assessment. For links, I considered the number of links constructed and whether they were arranged to form clear, concise, unambiguous propositions. For linking phrases, I checked how many links were labelled and whether such labels contained detailed information on the relationship. For organisation, I examined the legibility of the student-constructed maps and whether the students' mapping sequences were easy to follow.

Students' performance on the CMS-test suggests that the training effectively prepared most students to construct informative concept maps that reflect their understanding of the mathematical concepts. Moreover, these results served as a basis for further refinement of the training programme, specifically in the area of statement transformation. Over 75 percent of the students correctly expressed the given statements in the form of propositions. Difficulty arose when the given statement involved three or more concepts, especially when the students extracted all the involved concepts as nodes and combined them to form a single proposition. This suggests that additional training on statement transformation for more than two concepts is in order. During simple free association and extended free association tasks, the students included all the given concepts as nodes in propositions. Over 80 percent of the links were unidirectional and the other 20 percent were either not directed (—) or bidirectional (i.e. ↔). Some 95 percent of the links were labelled, 66 percent with labelled with detailed linking phrases. Linking phrases such as 'include', 'is', 'has', and 'may be' are deemed not detailed enough. That being said, except for two students, all participants included detailed linking phrases during the *simple free association* skill testing task and were thus apparently cognisant of the detailed linking phrases requirement. Most of the maps students constructed during the free association testing task were easy to follow. Almost 80 percent of the students paid attention to the hierarchy of the concepts even though they were not required to do so.

In summary, the students presented the concepts in node form, used links to indicate connections, and included linking phrases to indicate relationships. They were aware that the links should be unidirectional and that the linking phrases should be detailed. More in-depth practice may be needed for students to develop more complex statement transformation skills, and the requirement for detailed linking phrases should be further stressed to address some of the problems reported earlier.

Revised Training in the Main Study

This section reports on the finalised training programme for concept mapping in mathematics which was used in the main study.

Participants

The participants in the main study were from a class of Grade 8 students ($n = 48$, 24 females and 24 males, aged 13–14 years) in a junior middle school in Libao, a town in Jiangsu province, China. They indicated that they had no previous familiarity with concept map or concept mapping. After participating in the skills training and successfully acquiring the needed skills, they participated in the main study on concept mapping, the components of which are presented in Chapters 5, 6, and 7.

Training

Four 40-minute training sessions were conducted as previously described. Based on the results of the pilot study, statements with more than two key concepts were addressed, and the use of detailed linking phrases was further emphasised. Additional examples were included.

After I introduced the key attributes of concept map, I presented each student with a copy of a sample concept map on *numbers*. I asked them to read the concept map in preparation for learning to construct their own maps. This was intended to give students an overall impression of what a concept map is and how to use it to express ideas, specifying the general requirements for nodes, links, and linking phrases. Counter-examples were provided so that the students could better understand why concept names, rather than, e.g. phrases, should be used as nodes; why links should be unidirectional; and why linking phrases should be detailed (informative). The counter-examples were intended to assist students in distinguishing between well-constructed and poorly constructed concept maps. Students were informed that well-constructed concept maps should include rich and substantial links and include detailed and accurate linking phrases. The maps should be neatly organised so that the information conveyed is clear and explicit to others. In post-study interviews (see Chapter 7) students indicated that they had understood these criteria and articulated the relevant characteristics. For example, over 56 percent of students interviewed explained that the linking phrases of a good concept map should be accurate, detailed, complete, and appropriate.

More challenging examples which involved more than two key concepts were included in the training on *statement transformation*. Students were reminded that each proposition should convey one independent idea. I advised them to, when working with transforming statements involving three or more concepts, extract two key concepts as nodes and include the other concepts in the linking phrases or separate the ideas in the given statement into separate parts and construct propositions correspondingly. For example, the statement 'a positive number has two square roots which are opposite numbers' has three key concepts, positive number, square root, and opposite number. It can be expressed in two ways, as illustrated in Figure 4.7. In the proposition on the left side, *positive number* and *square root* are the nodes and the concept *opposite number* is included in the linking phrase. When all three concepts are used as nodes, the statement should be split into two ideas 'a positive number has two square roots' and 'the square roots of a positive numbers are opposite numbers'. Otherwise, when more concepts and propositions are included, ideas may be misunderstood. In the example presented in Figure 4.8, the two propositions on the left side together convey the idea of the given statement; however, when the concepts zero and negative number are added, the idea

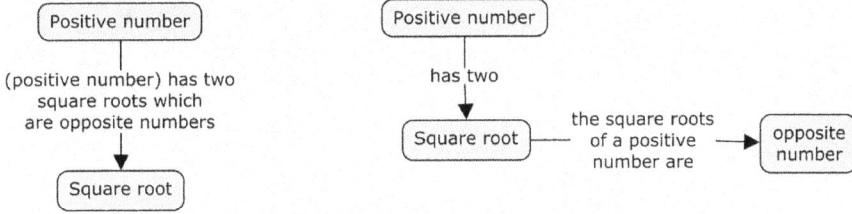

Figure 4.7 Examples of statement transformation with three key concepts.

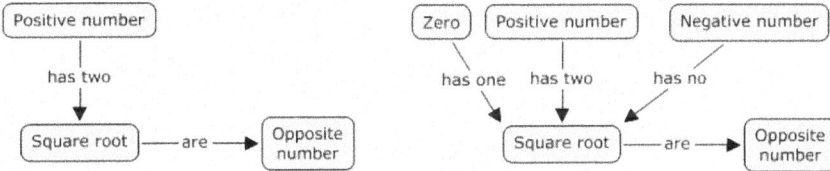

Figure 4.8 Counter-examples of statement transformation with three key concepts.

of the proposition 'square root(s) are opposite numbers' becomes blurred (see the propositions on the right side).

I assigned simple free association skills building exercises in class and observed students at work. Problematic associations were discussed in class. Through this discussion, the requirement that detailed linking phrases be included was further emphasised. In addition, I presented them with poorly constructed concept maps (see Figure 4.9) in our work on developing extended free associations skills. Students commented on the erroneous concept maps and offered suggestions for refining them. The concept maps can be refined by considering the following principles:

Closely related concepts should be in close proximity on the map.

In arranging the concepts, ensure that enough space is left between linked concepts to add linking phrases.

Use unidirectional links.

Avoid cross links to avoid confusing presentation.

Label all the links with linking phrases.

Use linking phrases that are sufficiently detailed that the relationships will be clear to others.

Add more connections, if possible.

Ensure that the entire map is clear and legible. Redraw the map if necessary.

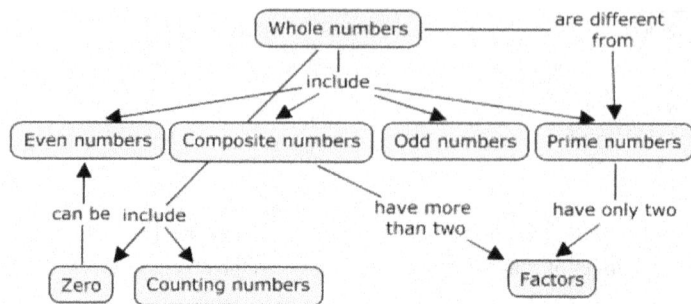

Figure 4.9 An example of a poorly constructed concept map.

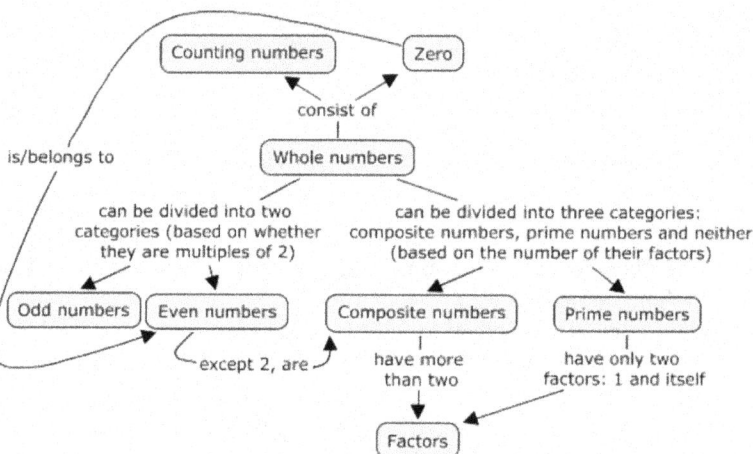

Figure 4.10 An example of a well-constructed concept map.

The concepts composite numbers and prime numbers are closely related and so should be arranged together. The links from whole numbers to zero and to counting numbers are crossed with the link from whole numbers to even numbers. This can be rectified either by moving the concept counting numbers to a more appropriate position or by joining the concepts whole numbers and counting numbers with a curved link. The linking phrases, e.g. 'include', in the map are too general. More information should be included in the linking phrases. In addition, more relationships can be included, e.g. '*even numbers*, except number two, are *composite numbers*'. In consideration of these principles, the concept map in Figure 4.9 can be refined into the concept map presented in Figure 4.10.

Results of post-training CMS test

A post-training CMS test (see Appendix A) was administered to assess students' mapping skills. As in the pilot study, evaluation focused primarily on the concepts, links, and linking phrases comprising the student-constructed propositions, emphasising the relevant skills.

> *Statement transformation:* A focus on the meaning of given statements after transformation and also the format of the propositions, i.e. nodes, unidirectional links, and linking phrases.
>
> *Simple free association:* A focus on the use of given concepts, format of propositions constructed, and detailed linking phrases.
>
> *Extended free association:* A focus on the use of given concepts, format of propositions, detailed linking phrases, and organisation of propositions.

Concept mapping skills aside, students' performance on the CMS test was likely also affected by their level of conceptual understanding of the material. For example, it is possible that a student could not provide a detailed linking phrase to a proposition because they did not know much about the relationship, even though they understood the requirements for building detailed linking phrases. Accordingly, students' performance on the CMS-tests in the main study was more systematically analysed than in the pilot study. In analysing the pilot study, I concentrated on concept mapping skills at the whole-class level in attempt to gain a holistic picture of the effectiveness of the training while still considering students' concept mapping skills individually. However, scoring students' CMS performance by directly counting the numbers of correct concepts, unidirectional links, labelled links, and the like, was found inappropriate. For example, consider *statement transformation*. Students were able to construct a correct proposition for the statement 'two numbers with different symbols but same absolute value are called opposite numbers' with either two concepts (e.g. absolute value and opposite numbers) or three concepts (symbol, absolute value, and opposite numbers). A larger number of correct concepts extracted from a given statement does not necessarily suggest better performance. Accordingly, I evaluated students' performance on the CMS-tests by summing the percentages of acceptable answers (see later). These scores were used with the students' concept map scores (see Chapter 6) and their expressed attitudes toward concept mapping for the correlational analysis reported on in Chapter 7.

The percentages in the table are calculated in the following manner:

For Statement Transformation

CONCEPT: the number of concepts correctly extracted from the given statements against the number of nodes used in the propositions.

LINK: the number of unidirectional links against the total number of links.

LABELLED LINKS: the number of labelled links against the total number of links.

MEANING: the number of propositions which shared the same idea as their corresponding statements against the number of propositions constructed.

For Simple Free Association and Extended Free Association

CONCEPT: the number of involved given concepts against the total number of given concepts.

LINKS: the same as for Statement Transformation.

LABELLED LINKS: the same as for Statement Transformation.

DETAILED LINKING PHRASES: the number of links with detailed linking phrases against the number of labelled links. Linking phrases such as 'include', 'is', 'has', and 'may be' are considered not detailed.

Organisation (for extended free association only): The organisation of the student-constructed concept maps is scored 1, 2, or 3, corresponding to whether the given concepts are weakly structured (1 point), partially structured (2 points), or hierarchically structured (3 points) as compared with a criterion map constructed by the researcher. A score of 0 is assigned for organisation if no links were built among the given concepts. For consistency, the percentage of organisation is calculated using the organisation score divided by the full mark of 3.

The maximum score for statement transformation is 100 percent (for concept) + 100 percent (for links) + 100 percent (for labelled links) + 100 percent (for meaning) = 4. Similarly, the maximum scores for simple free association and extended free association are 4 and 5, respectively. Accordingly, students' CMS scores could range from 0 to 4+4+5=13, where 0 indicates no response or totally irrelevant answers and 13 indicates fully correct responses. The last column in the table reports the percentages of propositions which are all correct with respect to the concepts, links, labelled links, detailed linking phrases, meaning, and organization.

For statement transformation, 90 percent of the nodes in the propositions are concepts correctly extracted from the given statements; 98 percent of the links are arrowed, the remaining 2 percent are not directed, and none are bi-directional. All the links are labelled. More than 50 percent of the linking phrases are copied directly from the given statements. This is acceptable given the requirements set in the training. Some 85 percent of the student-constructed propositions accurately expressed the meaning of their corresponding statements. In 12 percent of the propositions, the meaning was partially changed, and in 3 percent a completely different idea was provided. The percentage of acceptable meanings for the statement transformation is 91 percent.

For simple free association, students construct propositions using the given concepts. About 2 percent of the concepts are not included in some students' maps, though some related concepts are added. Concerning the links, 97 percent are unidirectional, 99 percent, regardless of whether they were arrowed correctly, are labelled, and 85 percent of the linking phrases include sufficient details. This percentage is acceptable because the students' ability to provide detailed linking phrases may have been limited by their understanding of the given concepts. Regarding the requirements set for this task, 73 percent of the propositions are correctly constructed. This percentage is slightly better than that in the statement

Table 4.1 Percentages of acceptable answers in the post-training CMS with respect to the requirements set for the tasks.

Task	Concept	Link	Labelled Links	Detailed Linking Phrases	Meaning	Organization	Fully Correct
Statement transformation	90%	98%	100%	/	91%	/	69%
Simple free association	98%	97%	99%	85%	/	/	73%
Extended free association	88%	98%	97%	89%	/	74%	54%

transformation (69 percent), suggesting that students may find simple free association easier than statement transformation.

The extended free association seemed to be the most difficult among the three task types. Three students did not attempt it. Among those who did, all knew that the given concepts were to be represented as nodes, and 81 percent of those students used all six concepts. Some 98 percent of the links constructed are correctly arrowed, and almost all the links are labelled with linking phrases. Almost 90 percent of linking phrases include detailed information. This finding suggests that most students knew that a detailed linking phrase is required for quality propositions. In addition, most of the students paid attention to the organisation of concepts in their concept maps. Over 50 percent of the students clearly presented the conceptual hierarchy, even though it was not required. Twenty-six of the 48 (54 percent) earned full marks for this task.

These findings are consistent with the students' claims in interviews and in response to open-ended questions (see Chapter 7). The students indicated that the extended free association task was the most difficult of the three task types. Students obtained the lowest percentage of fully correct answers (54 percent in Table 4.1) in that section of the CMS test. Although they reported that the simple free association was more difficult than the statement transformation because it required them to think hard about the relationships themselves, they presented a higher percentage of fully correct answers during the simple free association task. Maintaining the meaning of a given statement during the statement transformation may more challenging than adding detailed linking phrases in the simple free association task.

Comparison of Post-Training CMS Test and Post-Practice CMS Test

A parallel post-practice CMS test (see Appendix A) was administered after the students engaged in concept mapping practice for about three weeks. Table 4.2 shows the percentages of students' acceptable answers in the post-practice CMS test regarding the requirements set for the three skills testing tasks.

Table 4.2 Percentages of acceptable answers in the post-practice CMS test with respect to requirements set for the tasks.

Task	Concept	Link	Labelled Links	Detailed Linking Phrases	Meaning	Organization	Fully Correct
Statement transformation	90%	99%	100%	/	94%	/	74%
Simple free association	99%	98%	99%	87%	/	/	81%
Extended free association	94%	98%	100%	98%	/	87%	71%

A comparison of the percentages in Table 4.1 and Table 4.2 indicates that the students performed better on the post-practice CMS test than on the post-training CMS test. Since they already performed quite well on the concept, link, and labelled links aspect in the post-training CMS test, with percentages of correct responses over 90 percent, significant improvement of these skills was not expected due to ceiling effect, even after a period of extended practice. In addition to percentages of fully correct answers for the three types of skills, obvious increases were presented in detailed linking phrases in the extended free association task (from 89 percent to 98 percent) and in organisation for the extended free association (from 74 percent to 87 percent). Paired-sample t tests were conducted to further investigate whether the changes in the students' concept mapping skills were significant. The results of the t tests are reported in Table 4.3.

The paired-samples t test on the overall scores of the post-training and post-practice CMS tests indicates statistically significant improvement in students' concept mapping skills (Pair 1 in Table 4.3, $p < .005$). The paired-samples t tests for the other pairs in the table further indicate that the major improvements were in students' performance on the *extended free association* ($p < .005$), i.e. detailed linking phrases ($p < .05$) and organisation ($p < .01$). No statistically significant improvement was found in students' performance on the statement transformation and simple free association tasks. Students indicated during the interview and in the answers to the open-ended questions (see Chapter 7) that statement transformation and simple free association were easier than extended free association; they may have refined the two mapping skills well after the training. Moreover, the means reported in Table 4.3 suggest that almost all the students' concept mapping skills had improved in almost all aspects. This is also consistent with participating students' responses in the ATCMQ (see Chapter 7), i.e. over 90 percent of the students agreed with the statement that they could construct better concept maps with additional training and familiarity.

Table 4.3 Result of paired-sample *t* tests of the students' concept mapping skills as measured by the post-training and post-practice CMS tests.

		Maximum possible score	Post-training CMS		Post-practice CMS		t	Sig. (2-tailed)
			Mean	SD	Mean	SD		
Statement Transformation	1 CMS scores	13	11.82	1.73	12.38	1.00	3.20	.002
	2 ST* score	4	3.80	.26	3.83	.29	1.00	.324
	3 Concept	1	.90	.17	.90	.13	.04	.967
	4 Link	1	.98	.06	.99	.07	.37	.714
	5 Labelled links	1	1.00	.00	1.00	.00	–	–
	6 Meaning	1	.91	.15	.94	.15	1.43	.159
Simple Free association	7 SFA* score	4	3.79	.28	3.83	.23	1.09	.281
	8 Concept	1	.98	.06	.99	.05	.89	.377
	9 Link	1	.97	.07	.98	.06	1.01	.318
	10 Labelled links	1	1.00	.03	.99	.03	-.74	.461
	11 Detailed linking phrases	1	.85	.23	.87	.21	.58	.566
Extended Free association	12 EFA* score	5	4.61	.50	4.83	.31	3.31	.002
	13 Concept	1	.88	.29	.94	.18	.88	.066
	14 Link	1	.98	.05	.98	.06	-.28	.784
	15 Labelled links	1	.97	.15	1.00	.00	1.24	.220
	16 Detailed linking phrases	1	.89	.25	.98	.09	2.36	.023
	17 Organization	1	.74	.32	.87	.25	3.07	.004

* The ST score, SFA score, and EFA score denoting respectively the sum score of the three CMSs

Summary

This chapter reports on the development of a concept mapping training programme that can effectively prepare students to construct concept maps and which can be used by educators to assess their students' conceptual understanding in mathematics. The approaches presented were explored through the preliminary and pilot studies, and lessons learned from them informed the training conducted with participants in the main study. The results of the post-training CMS testing indicate that the skills students developed during the finalised training programme had prepared them to successfully present the extent of their conceptual understanding during the CM tasks in the main study. The comparison of the students' post-training CMS test performance and post-practice CMS test performance indicates that concept mapping skills can be further improved with practice.

Note

The training methods and students' mapping performance explored in the preliminary study has been published as a journal paper: Jin, H., and Wong, K. Y. (2010) 'Training on concept mapping skills in geometry'. *Journal of Mathematics Education*, 3(1), pp. 103–118.

5 Assessing Conceptual Understanding in Mathematics With Concept Mapping

Study Design

In Chapter 3, I reviewed different formats of CM tasks, and established two main categories of tasks, high-directed and low-directed. After comparing the limitations and strengths of the two types, I designed tasks with a balance of high- and low-directness. For the mapping tasks, using given concept lists helps students maintain focus on a specific knowledge domain, and at the same time allows them to freely connect concepts and label the links with their own words.

Participants

The participants were the same 48 Chinese Grade 8 students who participated in the training programme described in Chapter 4 and were, according to the CMS-test results described, prepared to express the full extent of their conceptual knowledge during the main mapping tasks.

Concept Mapping Tasks (CM Tasks)

Four different mathematical topics, two algebraic topics (algebraic expressions and equations) and two geometric topics (triangles and quadrilaterals) are explored to identify topic-specific issues in using concept mapping to assess conceptual understanding. The topics were chosen because they are the basic building blocks of students' mathematics learning and the most frequently included topics in mathematics curricula (e.g. Ministry of Education, China, 2011; Ministry of Education, Singapore, 2019; Ubuz Aydın, 2018; Weinberg et al., 2016; Yildiz et al., 2020). For each topic, I built a concept list from the participants' mathematics textbooks in consultation with two professors of mathematics education, one in China and one in Singapore. Before embarking on the CM tasks, the participants had learned the following concepts:

1   *Algebraic expressions*: integral expression, monomial, polynomial, fractional expression, coefficient, degree, constant term, like terms, unlike terms, common factor.

DOI: 10.4324/9781003269373-5

2   *Equations*: equation, solution, linear equation with one unknown, linear equation with two unknowns, $kx + b = 0$ ($k \neq 0$, k, b are constants), unknown, linear function, $y = kx + b$ (k, b are constants), proportional function, $ax + by + c = 0$ (a, b, c are constants).

3   *Triangles*: triangle, acute-angled triangle, right-angled triangle, obtuse-angled triangle, scalene triangle, isosceles triangle, equilateral triangle, axis of symmetry, angle, median, midline.

4   *Quadrilaterals*: quadrilateral, rectangle, parallelogram, square, rhombus, trapezium, isosceles trapezium, diagonal, centre of symmetry, axis of symmetry.

Data Collection

The four CM tasks for the topics were assigned during regular class time on different days within a two-week period. For each such task, participants had 30 minutes to construct a concept maps independently on size A4 paper. They were encouraged to include as many meaningful links as they could conceive of and to use complete and accurate linking phrases. During the data collection sessions, their mathematics teacher was present to make sure that they took the tasks seriously. The tasks were administered in Chinese and translated into English for reporting.

Data Analysis

A two-dimensional analysis of the student-constructed concept maps is here presented; maps are analysed along conceptual dimension and student dimension (see Table 5.1). The conceptual dimension involves individual concepts, pairs of

Table 5.1 Methods for the analysis of student-constructed concept maps.

		Student	
		Individual student	Whole class (48 students)
Concept	Individual concept	• number of incoming/outgoing links;	• mean number of incoming/outgoing links; • hierarchical position;
	Pairs of concepts (links associated with individual concepts)	• whether there is a connection between the pair of concepts; • whether the linking phrase of the connection is correct, partially correct, or wrong;	• how many students build a connection between the pair of concepts; • among the connections, how many are labelled with correct or partially correct linking phrases;
	Concepts in given list (the entire concept map)	• density; • proposition score	• hierarchy; • collective map;

concepts, and the concepts in the given list as a whole. The students dimension involves individual students and the whole class. To address the individual student dimension, examples of well-constructed and poorly constructed concept maps are presented together with the numerical scores awarded. This is also with the aim of determining the feasibility of concept mapping as a tool for investigating students' conceptual understanding in detail and in differentiating the quality of conceptual understanding among students. At the whole-class level, descriptive statistics are employed to summarise the students' conceptual understanding based on their individual concept maps.

The maps are not scored by comparison with a criteria map, because Ruiz-Primo and Shavelson (1996) found that different such maps may can lead to different conclusions. The key terms used in the analysis are as follows:

Number of Links

Concepts in concept maps are connected through unidirectional links. Taking the direction of the links into account, links to a concept can be categorised as incoming and outgoing. I count the incoming links, the outgoing links, and the total links (the sum of the incoming and outgoing link counts) to each given concept in a student's concept map. When a concept is isolated from others, it is considered to have zero incoming, outgoing, and total links.

The number of links between given concepts in a pair are also counted. In a student's concept map, two concepts, A and B, can be connected as 'A→B', or 'A←B', or not connected. No student used double-arrowed links (A↔B) or provided more than one link from one concept to another in this study. Thus, the number of links between concepts in a pair, e.g. A→B, is either one or zero in a concept map.

Density

Density measures the extent to which a group of concepts are connected in a concept map, calculated as the total number of constructed links divided by the number of concepts in a concept map. With the same number of concepts, a higher density suggests stronger connections among the concepts. In the present study, the concepts were given by the researcher; hence, for each concept mapping task, the number of concepts was fixed among the students' concept maps.

Compared with directly counting the number of links constructed, measuring density allows examiners to compare students' performance in different CM tasks, especially when the tasks have unequal numbers of given concepts. For example, students may construct more links for a concept mapping task with 20 given concepts than for a concept mapping task with ten. Directly comparing the numbers of links constructed for the two tasks could be misleading, because more links do not necessarily suggest stronger connections or better understanding. In this study, the concept mapping task for triangles includes a different number of given concepts than the other three CM tasks.

Hierarchy

The definition of hierarchy applied here is based on the conceptual relationships described by Pine (1985). According to Pine (1985), there are two fundamental types of conceptual relationships: set-subset relationships and whole-part relationships. Set-subset relationships are used in classification systems which can be formed on the basis of certain properties, attributes, or characteristics. Whole-part relationships refer to the organisation of structures and systems. Concepts considered *sets* or *wholes* are called superordinate concepts, and their corresponding *subsets* or *parts* are called subordinate concepts. Concepts, within a topic domain, that are neither in a set-subset relationship nor a whole-part relationship are called coordinate concepts. For example, triangles can be categorised as acute-angled, right-angled, and obtuse-angled according to the properties of its angles. Triangle and the three types of triangles are in a set-subset relationship, where triangle is the superordinate concept, and the three special types of triangles are subordinate concepts. Triangle and angle are in a whole-part relationship; hence, angle is also a subordinate concept of triangle. Acute-angled triangle, right-angled triangle, and obtuse-angled triangles are coordinate concepts because there is no set-subset or whole-part relationship among them.

Propositions

As emphasised in the training stage, detailed linking phrases are important for the assessment of students' conceptual understanding. Here, the proposition score of a student-constructed concept map is an assessment of the quality of the linking phrases. Each proposition is scored on a scale of 0 to 2. Proposition with an incorrect linking phrase scores 0 points, correct but incomplete linking phrase scores 1, and detailed linking phrase scores 2. The proposition score of a concept map is the sum of the points earned for all propositions in the concept map. The face validity of the coding and scoring of the propositions was confirmed in consultation with an associate professor of mathematics education at a university in Nanjing, China. The inter-rater reliabilities of the density and proposition scores were checked. A high school mathematics teacher was invited to be a rater. I spent about 30 minutes introducing her to the definition of density and the coding and scoring rubric for propositions with examples. The rater complete the scoring outside of the presence of the researcher at her convenience. The inter-rater reliability coefficients for the scoring of density were 1 for all the four topics and over 0.90 for propositions (algebraic expressions: 0.91, equations: 0.92, triangles: 0.94, and quadrilaterals: 0.96). This suggests that the scoring scheme can provide reliable information about the quality of students' concept maps.

Concept Mapping on Algebraic Expressions

The CM tasks in this study were free-style mapping with a list of the given concepts. For the topic algebraic expressions, ten relevant concepts were selected from the students' textbooks. The students had 30 minutes to construct a concept map using the given concepts. They were allowed to include additional related concepts. However, in analysing the concept maps, only the given concepts are considered so as to have a common basis for comparison of students' conceptual understanding and for generalisation to the collective map at the whole-class level.

Individual Student Concept Maps (Algebraic Expressions)

This section presents two examples of student-constructed concept maps: a well-constructed concept map and a poorly constructed concept map. It should be noticed that 'well-constructed' and 'poorly constructed' are relative terms, based on the number of given concepts included, the number of links constructed, and the quality of the linking phrases used. In this study on the concept map as an assessment and diagnostic tool, a poorly constructed concept map might indicate a moderate conceptual understanding.

The concept maps constructed were hand-drawn and in Chinese by the students. The concept maps presented in this chapter have been translated into English and redrawn using IHMC CmapTools software, available at http://cmap.ihmc. us. In redrawing, I tried to reproduce students' concept maps as closely as possible in terms of the relative positions of the concepts, the ways the links were constructed, and the linking phrases used.

Figure 5.1 is a sample well-constructed concept map on algebraic expression concepts. It includes the ten given concepts. No additional concepts were added. Integral expression, fractional expression, monomial, and polynomial expressions are hierarchically placed. Fractional expression, as a coordinate concept of

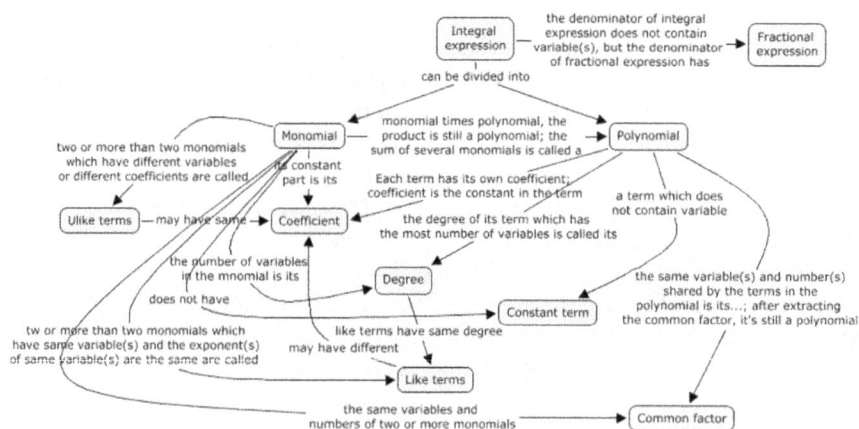

Figure 5.1 An example of a well-constructed concept map for algebraic expression concepts.

Note: Translated from Chinese

integral expression, is at almost the same level as integral expression. Monomial and polynomial are subordinate concepts of integral expression; they are at a lower level than integral expression. The other six concepts are subordinate concepts of monomial and polynomial; they are at a lower level than monomial and polynomial. These six concepts also seem to be randomly arranged on the map. This might be because except for like terms and unlike terms, which are coordinate concepts and supposed to be placed at the same level, no clear hierarchical relationship exists between them.

The student constructed 17 links among the concepts, for a *density* of $17/10 = 1.7$. Most links are labelled with detailed linking phrases. At the top of the map, the linking phrase in the proposition '*integral expression* can be divided into *monomial* and *polynomial*' is considered not as detailed, resulting in a proposition score of 1. An example of a detailed linking phrases for this proposition is '[*integral expression*] can be divided into *monomial* and *polynomial* according to the number of terms of the expression; if it has only one term, it is a *monomial* and if it has two or more terms, it is a *polynomial*'. This describes two categories of integral expressions.

The student recognised the difference between integral and fractional expressions. Instead of using the definition of fractional expression directly as the linking phrase, she states that the denominator of integral expression does not contain variable(s), but the denominator of fractional expression does. This indicates that she did not simply memorise the definition. She interpreted this by comparing the attribute integral expression with fractional expression.

The student found more than one relationship for some paired concepts. For example, *polynomial* and common factor were connected with both definition-based and property-based linking phrases, both being partially correct. The definition-based linking phrase states that the common factor of a polynomial is 'the same variable(s) and number(s) which are shared by the terms in the polynomial'. This is considered partially correct. A counter-example is that $3a$ is the common factor of $6a^2b + 9ac$; 'a' is the same variable, but 3 is not the same number shared by the two terms in the polynomial. The same number(s) should be the greatest common divisor of the numbers in the terms. The student may have been limited in her ability to describe ideas accurately. Moreover, the linking phrase 'after extracting common factor, it's still a polynomial' is considered incomplete, since it does not specify whether the two polynomials have same number of terms or not. In the case that a student includes two linking phrases connecting the same two concepts in a pair, each proposition is assessed and scored separately. This scoring method is reasonable because the student who provides more than one linking phrase for a connection may know more about the relationships between the connected concepts than the student who provides only one linking phrase. The connection between *polynomial* and *common factor* in this student's map earned a score of $1 + 1 = 2$.

In general, the propositions in this map together indicate that the student accurately understood the concepts and the relationships between them. The proposition score of this concept map is $2 + 1 + 1 + 2 + 2 + 2 + 2 + 1 + 1 + 2 + 2 + 2 + 2 + 2 +$

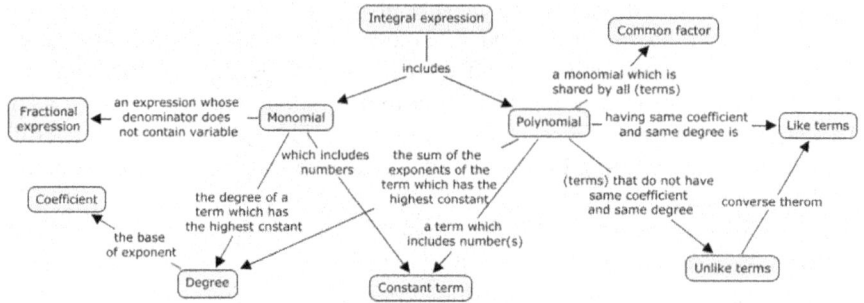

Figure 5.2 An example of a poorly constructed concept map for algebraic expression
concepts.

Notes: Translated from Chinese

$1+2+1=30$. This is one of the highest proposition scores for the concept maps
(*algebraic expression*).

Figure 5.2 is an example of a poorly constructed concept map (algebraic
expression). It includes all ten given concepts. However, only 12 links were constructed. Its density is $12/10=1.2$, lower than the density of the well-constructed
concept map (Figure 5.1). Except for *integral expression, monomial*, and *polynomial* which are arranged hierarchically, the hierarchies of the other concepts are
not clearly indicated. The student might not have considered the hierarchy of the
concept map, which is acceptable because the hierarchical organization was not
required.

Overall, the linking phrases indicate that the student may have had a poor
understanding of the concepts. Misconceptions are revealed. For example, the
*propositions from polynomial to like terms and unlike terms indicate that the stu-
dent thought that like terms* have the same coefficient and same degree and that
unlike terms do not have the same coefficient and same degree. Student's understanding of the relationships between some concepts in a pair of may have been
fuzzy. For example, the proposition '*like terms* converse theorem *unlike terms*'
indicates that she knew *that like terms and unlike terms are incompatible* since the
word *converse* suggests two concepts holding opposite ideas but could not state
it clearly and perhaps had not figure out which aspects of the two concepts are
incompatible. The proposition from polynomial to degree is almost correct except
for the use of the word *constant* instead of 'power' or 'degree'. The linking phrase
in this proposition is based on the definition of the degree of polynomials. The
student might have memorised the definition by rote.

This concept map earned a proposition score of $1+1+0+1+0+1+1+0+0+
0=5$. This score is quite low. Among the 11 propositions constructed, only five
reflected partially correct ideas, suggesting that the student did not have an accurate understanding of the relationships among the concepts.

Analyses of the concept maps (Figures 5.1 and 5.2) demonstrate the concept map's potential for assessing students' levels of understanding of the given concepts. The qualitative description of the concept maps yields information on students' knowledge about the relationships between the concepts, including misconceptions. The density and proposition scores, to a certain extent, reflect the quality of the students' conceptual understanding as assessed by the concept mapping task. The descriptive statistics of the densities and the proposition scores of the students' concept maps and the correlation between the two variables are reported in Table 5.2.

The density of the students' concept maps ranges from 0.20 to 1.80. A density of 0.2 indicates that only two links were built among the ten given concepts; at least six of the concepts were left isolated (unconnected) in the concept map. The students' proposition scores ranged from 2 to 30. The Spearman's correlation of 0.793 ($p < .001$) indicates that the two variables are highly correlated. This is reasonable because the density and the proposition score of a concept map are both related to the number of links in the map, i.e. the more links constructed, the higher the density and the greater the likelihood of a higher proposition score.

Links Associated With Individual Concepts (Algebraic Expressions)

The analyses of the student-constructed concept maps at the whole-class level focus on two aspects of conceptual understanding: individual concepts and relationships between concepts in a pair.

Individual Concepts (Algebraic Expressions)

The concept mapping task (algebraic expression) addresses students' knowledge of the individual concepts from the perspective of the links between them. The number of links connected to each of the concepts in the student-constructed concept maps reflects the extent to which the concept connects to the other concepts in the maps. A higher number of links indicates stronger connections with other concepts in the domain. Hence, the total number of links (sum of including and outgoing links) to a concept could be an indicator of the students' conceptual understanding.

Table 5.2 Descriptive statistics of density and proposition score of concept maps and their correlation (algebraic expressions).

	Minimum	Maximum	Mean	SD	Spearman's Correlation
CM density	0.20	01.80	01.21	0.34	.793**
CM proposition score	2.00	30.00	15.42	7.55	

** Correlation is significant at the .01 level (2-tailed)

Table 5.3 Mean number of links of individual concepts (algebraic expressions).

Concept	Mean number of incoming links (I)	Mean number of outgoing links (O)	Mean number of total links (I + O)	Percentage of acceptable links
Polynomial	1.76	3.73	5.49	67%
Monomial	1.04	3.66	4.70	68%
Integral expression	0.95	2.25	3.19	48%
Degree	2.27	0.28	2.55	77%
Like terms	1.18	1.37	2.55	67%
Coefficient	1.92	0.19	2.11	63%
Unlike terms	1.02	0.76	1.78	60%
Constant term	1.16	0.42	1.57	58%
Common factor	1.23	0.25	1.48	70%
Fractional expression	0.88	0.51	1.39	38%

Table 5.3 reports the mean numbers of links connected to each given concept in the 48 student-constructed concept maps. For ease of illustration, the concepts in Table 5.3 are sorted by the mean number of total links, in descending order, rather than as they are ordered in the given list. The percentage of acceptable links (the last column) for a concept is defined in such a way as to indicate students' overall understanding of the conceptual connections associated with that concept. Let T be the total number of links for a concept, of which C and P denote the number of links with correct or partially correct linking phrases, respectively. Using the scores of 2 or 1 for these links, respectively, the percentage of acceptable links is defined as $(2C + P)/2T$, expressed as a percentage. This is similar to the mean score expressed as a percentage of the maximum possible score if all the links had correct linking phrases. For example, for *integral expression*, 41 incoming links were constructed, among which only one was labelled with a correct linking phrase, 36 were labelled with partially correct linking phrase, and the other seven were labelled with incorrect linking phrases; 97 outgoing links were constructed, 12 of which were labelled with correct linking phrases, 71 were labelled with partially correct linking phrases, and the remaining four were labelled with incorrect linking phrases. If we ignore the directionality of the links, for *integral expression*, students constructed 41 + 97 = 138 links, among which 1 + 12 = 13 were labelled with correct linking phrase, 36 + 71 = 107 were labelled with partially correct linking phrases, and 7 + 4 = 11 were labelled with incorrect linking phrases. Then the percentage of acceptable links for *integral expression* is $(13 \times 2 + 107 \times 1)/(138 \times 2) \times 100$ percent = 48 percent. This percentage is relatively low because the majority of the linking phrases are partially correct.

The column for mean number of total links in Table 5.3 illustrates that students constructed more total links for polynomial, monomial, and integral expression than the other concepts, suggesting that these three concepts are the main 'actors' in the student-constructed concept maps. They feature very strong connections with the other concepts. The mean numbers of outgoing links column illustrate that most of the links are outgoing links from the three concepts and that the mean numbers of outgoing links for each of these three concepts is noticeably greater than for the other concepts. One possible reason for the phenomenon is that these three concepts are more inclusive than the other given concepts in the domain of algebraic expression. Moreover, it is noticed that students tended to construct whole-part relationships from those three concepts to the more specific concepts, i.e. coefficient, degree, and constant terms. The percentage of acceptable links column illustrates that students constructed a noticeably lower percentage of acceptable links with integral expression than with monomial and polynomial. This is because many of the links with in*tegral expression* are labelled 'includes (*monomial* and *polynomial*)' or 'can be divided into (*monomial* and *polynomial*)'. Such linking phrases are not considered to contain the degree of detail emphasised in the training; as such, they are considered partially correct and score one point.

Like terms and unlike terms are coordinate concepts. They are defined based on the variables in these terms and the exponents of the variables. The mean number of total links to these two concepts, together with the percentages of acceptable links, suggests that the students were more familiar with like terms than unlike terms. This is understandable because in order to decide whether two terms are like or unlike, one only need refer to the definition of like terms. If the given terms do not satisfy the conditions for like terms, they can be classified as unlike terms on that basis. Accordingly, the students used the definition of like terms more often than that of unlike terms.

Fractional expression has the lowest mean number of total links with the other concepts and the lowest percentage of acceptable links. In the mathematics curriculum in China, secondary students are only required to learn the definition of fractional expression and perform simple calculations. The coefficient, degree, and other aspects of fractional expressions are not discussed at the secondary level. This might be why few links were constructed with fractional expression. Some students confused fractional expressions with fractions. Other students seemed to remember the definition of fractional expression but could not accurately express the idea in words and did not provide correct linking phrases for the relevant links, which resulted in a low percentage of acceptable links.

Students constructed relatively fewer links from common factor to the other concepts, but they constructed the second highest percentage of acceptable links from this concept. This suggests that, mathematically, it may have few links with other concepts. In basic algebra, common factor is a term used in factorisation, which aims to reduce numbers to relatively prime numbers or polynomials to irreducible polynomials. In the knowledge structure possessed by these students, common factor may be more related to procedural steps than to conceptual relationships.

Relationships Between Concepts in a Pair (Algebraic Expressions)

Concept mapping can directly address relationships between concepts in a pair. This section reports on the findings from the concept mapping task (algebraic expression) on students' knowledge of the pairwise relationships between the given concepts.

Table 5.4 reports the number of connections the 48 students constructed between concepts in a pair. Since each student constructed no more than one link between each concept in a pair, the numbers indicate students' degree of familiarity with the relationship between concepts in pairs. The number before the slash in the pairs of numbers presented in the table (e.g. 30/30) is the total number of connections students perceived between two concepts, regardless of whether the connection is valid or correctly expressed. The number after the slash represents the number of valid or correct expressed or partially correctly expressed connections. This indicates the degree to which students understood the connection. For example, '14/9' for unlike terms to degree indicates that 14 out of the 48 students linked unlike terms to degree and that nine such links were labelled with either correct or partially correct linking phrases and that five of the links were either not labelled or labelled with incorrect linking phrases. A zero in a cell indicates that none of the students linked the two terms. The last row and the last column in the table summarise the number of incoming and outgoing links, respectively, for each given concept in the 48 student-constructed concept maps; divided by 48, the numbers before the slashes are the same mean numbers of incoming links and the mean number of outgoing links, respectively, reported in Table 5.3. The sum of the numbers of links, together with the sum of the correct or partially correct links, provides information about the overall rates of correctly expressed connections between concepts in a pair. Finally, '579/494' in the last cell in Table 5.4 indicates that among the 579 links constructed for this topic, 494 (85 percent) were labelled with correct or partially correct linking phrases, suggesting that the students were generally knowledgeable about the relationships they constructed.

To outline the students' cognitive structure as a class, I constructed a collective map (see Figure 5.3) by considering link strength (total number of links constructed; see the first numbers in the pairs of numbers in Table 5.4). The map offers a rough but holistic picture of the connections the students' illustrated in their concept maps.

In the collective map, the strengths of the connections are expressed by three different types of arrowed lines, a thicker line, a thinner line, and a dotted line. The thicker arrows represent cases where more than 24 students (50 percent) built links between the concepts in a pair. The thinner arrows indicate that less than 24 (50 percent) but more than 12 students (25 percent) linked the concepts in a pair. The remaining paired concepts, those linked by more than zero but less than 12 students, are connected by dotted arrows. The same key is used for the three other collective maps presented in this study.

In Figure 5.3, the relative nodes sizes represent the mean numbers of total links with them (see Table 5.3). *Polynomial* is largest node, because it was linked with the greatest number of other concepts; accordingly, monomial and integral expressions, linked with fewer concepts, are smaller still, and common factor (linked with the least number of concepts) is the smallest node.

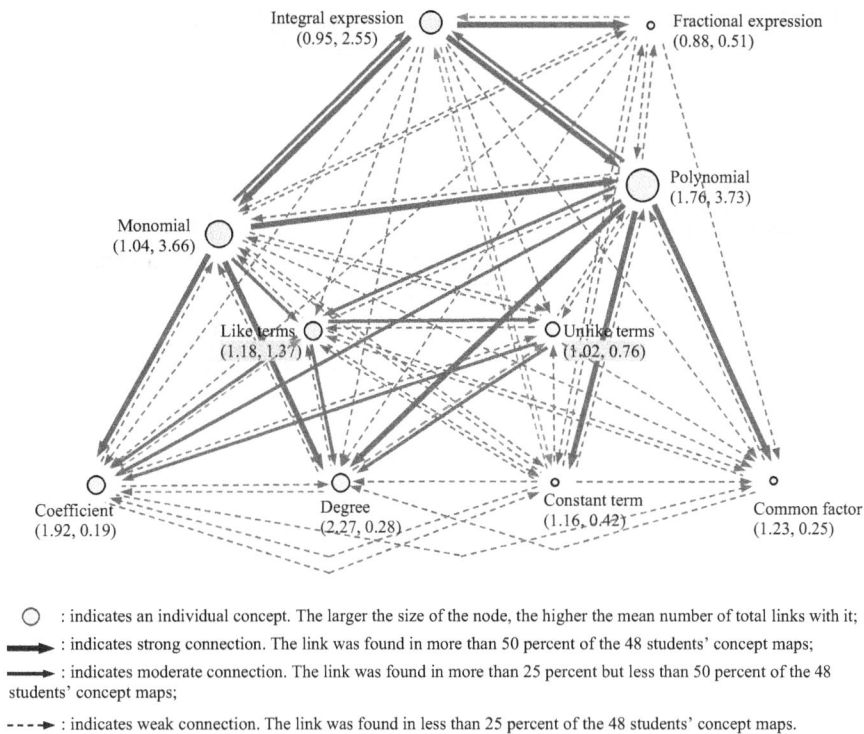

Figure 5.3 Collective map showing the connections between pairs of algebraic expression concepts.

In the collective map, conceptual hierarchies are differentiated according to superordinate, coordinate, and subordinate relationships as indicated in the concept definitions in the students' textbook. For example, integral expression is the most inclusive concept among the concepts, and so is placed at the top of the map. Fractional expression, a coordinate concept of integral expression, is at about the same level as integral expression. Coefficient, which is subordinate to most of the concepts in the given list, is at the bottom level of the collective map together with degree, common factor, and constant term.

The collective map illustrates that the strongest links are top-down, except the link from integral expression to fractional expression and the link from monomial to polynomial. This suggests that most students tended to link from more general or more inclusive concepts to more specific or less inclusive concepts.

Integral expression and fractional expression are two coordinate concepts. Fractional expression is introduced after integral expression in the students' textbook. This might be why most students linked integral expression to fractional expression. A notable finding from the data presented Table 5.4 is that, with the exception of the links from integral expression to fractional expression, all the

Table 5.4 Number of links constructed by the 48 students for each pair of concepts (algebraic expressions).

From \ To	Polynomial	Monomial	Integral expression	Degree	Like terms	Coefficient	Unlike terms	Constant term	Common factor	Fractional expression	Total outgoing links from
Polynomial		6/6	17/16	**32/28**	**19/14**	**15/13**	11/8	**31/29**	**28/25**	2/0	161/139
Monomial	**30/30**		16/15	28/26	14/14	**32/28**	10/9	11/11	12/12	5/0	158/145
Integral expression	**27/26**	**27/25**		1/1	3/2	1/1	2/0	4/2	2/2	**30/24**	97/83
Degree	0	1/1	0		5/5	2/0	4/4	0	0	0	12/10
Like terms	4/2	1/1	0	**16/16**		**16/10**	**14/11**	1/1	7/7	0	59/48
Coefficient	0	1/1	0	2/1	1/1		2/2	1/0	1/0	0	8/5
Unlike terms	2/2	1/1	0	**14/9**	2/2	**13/12**		0	1/0	0	33/26
Constant term	3/3	2/2	2/1	2/2	5/3	1/1	1/0		1/1	1/0	18/13
Common factor	7/7	0	0	3/2	1/1	0	0	0		0	11/10
Fractional expression	3/2	6/3	6/5	0	1/1	3/2	0	2/1	1/1		22/15
Total incoming links to	76/72	45/40	41/37	98/85	51/43	83/67	44/34	50/44	53/48	38/24	579/494

other links with fractional expression were labelled with invalid linking phrases. This may indicate that students' conceptual understanding of fractional expression was incorrect; incorrect understanding of an individual concept can obviously result in incorrectly presented connections with other concepts.

Integral expression is accurately connected with monomial and polynomial expressions. Students tended to link either from integral expression to monomial or polynomial or vice versa. From integral expression to monomial or polynomial, most students used the linking phrase 'includes' or 'can be divided into' to indicate the set-subset relationship between the concepts. The definition of *integral expression* may be the reason that a moderate number of students linked monomial and polynomial to integral expression. In the students' textbook, integral expression is defined as follows: '*monomial* and *polynomial* together are called *integral expressions*'. Accordingly, the students may have chosen to link the concepts in the order presented in the definition, that is, '*monomial* + *polynomial* → *integral expression*'.

The textbook defines polynomial as 'the sum of several monomials'. Hence, *a* monomial can be considered part of *a* polynomial. The whole-part relationship indicates that polynomial is higher in the hierarchy than monomial. However, a majority of the students (30 out of the 48) constructed a link from monomial to polynomial rather than from polynomial to monomial; and most of those students used the linking phrase 'the sum of several monomials is called', which is definition-based linking phrase. The order in which the concepts appear in the definition might be why they did not link from the more inclusive concept polynomial to the less inclusive concept monomial.

When only one or two students link two concepts, a misconception in those students' understanding is likely. For example, two students linked integral expression to unlike terms. It is possible that they had deeper insight into the relationship between the two concepts; however, the linking phrases indicate that they in fact understood the relationship wrongly. From this perspective, consideration of the correctness of the propositions can help eliminate insubstantial links and reveal weaknesses in their students' understanding.

During the concept mapping, after students had worked for about 25 minutes, and most students had built up their maps, I presented the following prompts to guide them toward more complete expression of what they knew about the given concepts and the relationships between them. The first two prompts were selected because students incorrectly defined like terms and unlike terms in the definition-example-nonexample task (DEN task) (algebraic expression) (see Chapter 6). The third prompt is related to Item 9(4) in the paper-and-pencil (P&P task) (algebraic expression) (see Chapter 6). About 25 percent of the students answered this item incorrectly, and few students had linked polynomial to common factor in their concept maps.

1   What is the relationship of the coefficients of like terms? How about degree of like terms?

2   What is the relationship of the coefficients of unlike terms? How about the degrees of unlike terms?

3   Will the number of terms of a polynomial be changed after extracting a common factor?

With these prompts, I observed some students adding connections to their concept maps. For their convenience and due to the time limit, I did not ask them to mark the newly added connections in different colours from the rest of the map to distinguish links added after the prompts and so am unable to determine which propositions were added as a result of the prompts.

However, even after having received the prompts, only a third of the 48 students linked like terms to coefficient. This suggests that the majority of students were unclear on the relationship between these two concepts. Among the links constructed, six were labelled with inaccurate linking phrases, such as by stating that like terms have the same coefficient. Another third of the 48 students linked like terms to degree. These links were all labelled with correct linking phrases, suggesting that they were clear that like terms have the same degree. Approximately 27 percent of the 48 students constructed a link from unlike terms to coefficient. The propositions formed were mostly correct (with one exception) and indicated that unlike terms may have same coefficient. Links from unlike terms to degree were also found in 14 out of 48 students' concept maps. Five students labelled the link with incorrect linking phrases, indicating that they thought that unlike terms should have different *degrees*.

With the prompts, nearly 60 percent of the 48 students linked polynomial to common factor. This is more than twice the number of students who constructed the link from monomial to common factor. Students used both definition-based and property-based linking phrases in the links from polynomial to common factor, while they primarily used definition-based linking phrases for the links from monomial to common factor. Two students indicated an incorrect understanding of the relationship. One of them thought that the number of terms of a polynomial changes after extracting a common factor; another defined common factor as the common number(s) of the terms in a polynomial. Eight students drafted partially correct linking phrases, indicating that after extracting a common factor, there are no changes to the polynomial. This proposition indicates a vague understanding of the relationship. More information is needed from the student to clarify the precise meaning of the proposition.

Analysis of the concept maps at the whole-class level suggests students' understanding of the given concepts in general. The mean numbers of links between the concepts together with the collective map facilitate identification of the key concepts of the given domain in the students' knowledge structure. They reflect the extent to which a concept is connected with other concepts in the domain, according to the students. From another perspective, the percentage of acceptable links calculated for each concept reflects students' understanding of the concepts. A higher percentage of acceptable links suggests a better understanding of the links with the concept. The number of links, together with the linking phrases constructed between pairs of concepts, indicates how strongly the concepts are connected, according to the students. Strong connections between certain concepts are observed. Some of these connections are due to students' frequent exposure to particular exercises. Analysis of the links and corresponding linking phrases with fractional expression suggests that students' misunderstanding or superficial understanding of a concept may result in wrong expressed connections with the concepts and that the method is thus suitable for revealing areas that require additional clarification for students.

Concept Mapping on Equations

For the topic equations, a list of ten concepts was provided to the students in the following order (translated from Chinese): equation, solution, linear equation in one unknown, linear equation in two unknowns, $kx + b = 0$ (k, b are constants), unknown, linear function, $y = kx + b$ (k, b are constants), proportional function, and $ax + by + c = 0$ (a, b, and c are constants). The students had 30 minutes to construct a concept map using the given concepts.

Individual Student Concept Maps (Equations)

Two student-constructed concept maps are presented in this section to illustrate students' conceptual understanding of equations. A well-constructed concept map is presented in Figure 5.4.

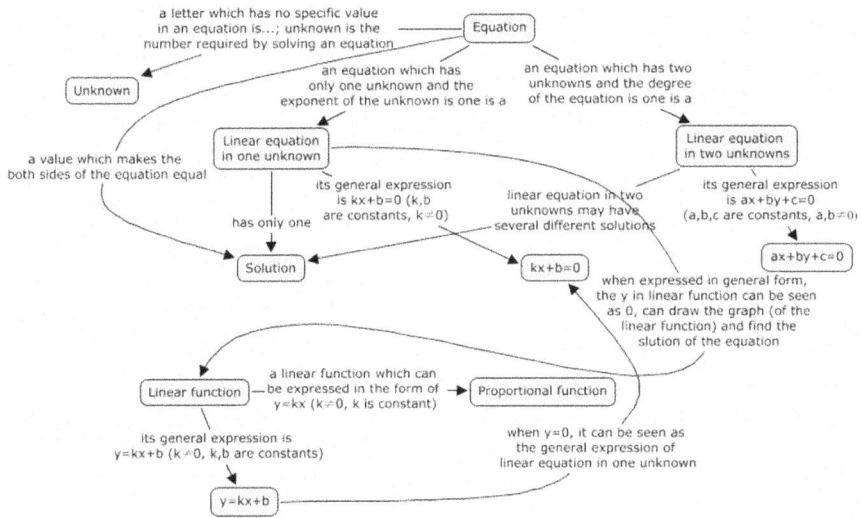

Figure 5.4 An example of a well-constructed concept map for equation concepts.

Source: Jin and Wong, 2015, p. 692

The concept map in Figure 5.4 covers the ten given concepts. No additional concepts were added. The concepts are generally divided into two clusters: the equation cluster in the upper half of the map and the function cluster in the lower half of the map. The two clusters are connected through linear equation in one unknown and linear function and their corresponding general expressions. The student recognised that when the y in the linear function $y = kx + b$ equals zero, the function can be considered a linear equation in one unknown $kx + b = 0$, and the solution of the equation can be found with the help of the linear function graph. Equation was taught to the students in the first semester of Grade 7, and linear function was taught in the first semester of Grade 8. No explicit connection between the two concepts was made in either the Grade 7 or Grade 8 textbooks. The connections between the two clusters in this student's concept maps are insightful.

The student's mapping sequence is easy to follow. The most inclusive concept equation is at the top of the map; it is directly connected to linear equation in one unknown, linear equation in two unknowns, unknown, and solution. Linear equation in one unknown and linear equation in two unknowns are two special types of equations; unknown is a part of an equation; solution is an unknown value that makes both sides of an equation equal. The student had a thorough understand of the hierarchical relationships between equation and these four concepts. The numerical expressions, i.e. $kx + b = 0$ and $ax + by + c = 0$, were correctly connected with the corresponding word representations. Linear function, proportional function, and $y = kx + b$ were placed together. The student recognised that proportional function is a linear function; he even pointed out in the linking phrase that it can be expressed in the form of $y = kx$ ($k \neq 0$ and k is a constant).

This concept map earned a density of $12/10 = 1.2$ and a proposition score of $2+2 + 2+2 + 2+2 + 2+2 + 2+2 + 2+2 = 24$. Although neither its density nor its proposition scores are among the highest, all of the propositions in this map score two points, since they are correct, and some even indicate insights into the relationships. Accordingly, it is considered a well-constructed concept map.

Figure 5.5 is an example of poorly constructed concept map for the equation concepts. Few links were constructed. Some linking phrases indicate a misunderstanding of the concepts.

This student omitted one given concept $ax + by + c = 0$ but added a relevant concept $y = kx$ The concepts are not hierarchically arranged like the map in Figure 5.4. Unknown is at the top of the concept map. It is directly linked to equation and the two special types of equations. However, the links from unknown to the two special types of equations are labelled with incorrect linking phrases. The student might not have been clear on the definitions of the two equations. He mistook the number of unknowns in an equation as the coefficient of the unknown. The link from linear equation in one unknown to linear function and the link from equation to linear function both suggest that the student realised that there was a certain connection between equations and functions but was unable to clearly express the connection. The linking phrase 'can be expressed as' is vague. Such links are considered partially correct and awarded one point.

Unknown

when the coefficient of the unknown is 1 and (contains) variable

the coefficient of unknown is 2 and contains variable(s)

an expression which contains unknown(s) and an equal sign

Linear equation in one unknown

Linear equation in two unknowns

can be expressed as

Equation

can be expressed as

Linear fucntion

which can get by solving th eequation

when k≠0, k is a constant, it is a

is not necessarily a

Solution

$y=kx+b$

Proportional function

y=kx+b and kx+b=0, one is the general expression of linear function, one is an equation which can be used to solve inequility

when k≠0 and k is a constant, it is a

$y=kx$

$kx+b=0$

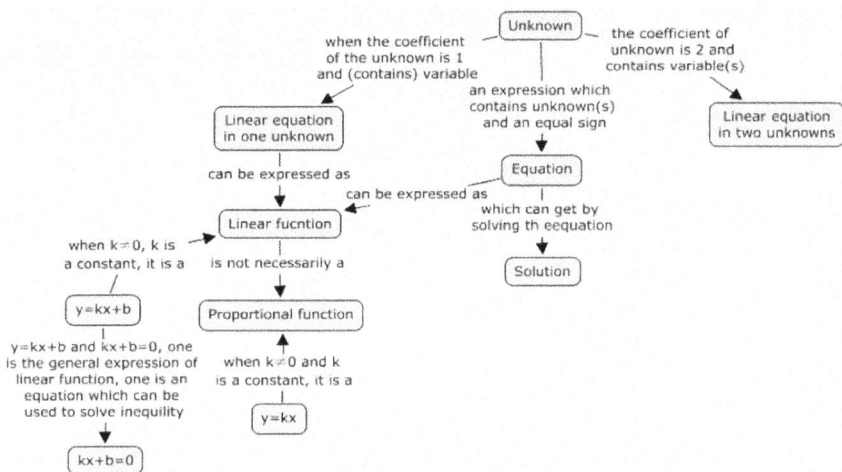

Figure 5.5 An example of a poorly constructed concept map for equation concepts.

Source: Jin and Wong, 2015, p. 693

Table 5.5 Descriptive statistics of density and proposition score of concept maps and their correlation (equations).

	Minimum	Maximum	Mean	SD	Spearman's Correlation
CM density	0.30	02.20	01.14	0.36	.734**
CM proposition score	0.00	31.00	15.94	6.97	

** Correlation is significant at the .01 level (2-tailed)

Only nine links were constructed among the nine given concepts in the concept map. The omitted concept $ax + by + c = 0$ is considered to be isolated. The link between the newly added concept $y = kx$ and proportional function is not scored. The density of this concept map is $9/10 = 0.9$, which is relatively low. The student might not have thought about further connections among the concepts as long as the concepts were connected. The proposition score is $0 + 2 + 0 + 1 + 1 + 2 + 2 + 2 + 2 = 12$, which is lower than the mean proposition score.

The descriptive statistics of the densities and the proposition scores of the students' concept maps and the correlation between the two variables are reported in Table 5.5.

The density of the students' concept maps of the equation concepts ranges from 0.30 to 2.20, indicating that all the students attempted the task. One student constructed 22 links among the ten given concepts. Another student constructed only three links, indicating that about half of the concepts were isolated in his concept map. The proposition scores of the students' concept maps range from zero to 31. Students' proposition scores vary greatly, as indicated by the large standard deviation (6.97). The Spearman's correlation of 0.734 ($p < .001$) indicates that the two variables are highly correlated.

Links by Individual Concept (Equations)

As with algebraic expressions, I analysed the student-constructed concept maps at the whole-class level on the following aspects of conceptual understanding by individual concepts and by relationships between concepts in a pair.

Individual Concepts (Equations)

Table 5.6 reports the mean numbers of links connected to each of the given concepts in the students' concept maps and the percentage of acceptable links for each concept. The concepts are presented in descending order of the mean number of total links.

Topping the list is equation. This is the most inclusive concept among the given concepts. The other concepts can be seen as either its parts or its subsets. Students tended to link it to its subordinate concepts (see the mean number of incoming links column and the mean number of outgoing links column in Table 5.6). Most of the links with equation are outgoing. The percentage of acceptable links with equation is 69 percent. Most of the links with equation were labelled with general linking phrases such as 'include' and 'can be divided into'. Such linking phrases earned a score of one point. Accordingly, the percentage of acceptable links for equation is low.

Table 5.6 Mean number of links of individual concepts (equations).

Concept	Mean number of incoming links (I)	Mean number of outgoing links (O)	Mean number of total links (I + O)	Percentage of acceptable links
Equation	1.00	3.02	4.02	69%
Linear equation in one unknown	1.42	1.73	3.15	72%
Linear equation in two unknowns	1.54	1.42	2.96	62%
Linear function	1.19	1.52	2.71	74%
Unknown	1.73	0.92	2.65	69%
Solution	1.56	0.31	1.87	65%
$y = kx + b$	0.85	0.79	1.64	77%
$kx + b = 0$	0.85	0.67	1.52	74%
Proportional function	0.69	0.69	1.38	60%
$ax + by + c = 0$	0.56	0.33	0.89	89%

Source: Jin and Wong, 2015, p. 694

Students had constructed roughly equal mean numbers of incoming and outgoing links to linear equations in one unknown and linear equation in two unknowns, special types of equations. The percentages of acceptable links for the two concepts suggest that the students knew more about linear equation in one unknown than about linear equation in two unknowns. They constructed relatively fewer links with $kx + b = 0$ and $ax + by + c = 0$ than with the corresponding word representations. Two possible reasons are as follows: First, students might have been less familiar with the numerical format. Second, they might have been better at making connections between concepts represented in the same formats than between concepts represented in different formats. Seven of the given concepts were represented in word format and three in numerical format. Students were able to construct more links with linear equation in one unknown and linear equation in two unknowns than with the numerically represented equivalents.

Linear function can be seen as a special type of equation with two variables. Participating students might not have had sufficient understanding of that relationship. Their concept maps presented few connections between linear function and equation. They were taught equations in Grade 7 and linear functions in Grade 8. Proportional function and $y = kx + b$ are the only two concepts in the given list that are obviously related to linear function. This might be a reason for the lower mean number of total links with linear function compared to equation concepts, i.e. equation, linear equation in one unknown, and linear equation in two unknowns. However, the percentages of acceptable links suggest that the students were aware of linkages with linear function but to a lesser degree than with equation concepts. Students constructed the lowest percentage of acceptable links for proportional function. This suggests that they held misconceptions about connections with proportional function or that they used only partially correct linking phrases. Among the 69 links with proportional function, 16 indicated misunderstandings and 23 indicated only partially correct ideas.

The expression $y = kx + b$ had a larger number of total links than the other two expressions, $kx + b = 0$ and $ax + by + c = 0$. Students might have been more familiar with the general expression for a linear function than with the general expressions for equations. There were more exercises in the students' textbooks dealing with $y = kx + b$. To correctly draw or interpret the graphical representation of linear function, students must be familiar with the expression, especially the meanings of k and b. Students generated high percentages of acceptable links for the three expressions. The majority of the links with these expressions were labelled with either correct or partially correct linking phrases. The students generally recognised that they were dealing with general expressions of linear function, linear equation in one unknown, and linear equation in two unknowns.

Relationships Between Concepts in a Pair (Equations)

This section reports on the students' knowledge of the relationships among the given concepts with respect to their performance in the concept mapping task for equations. Table 5.7 presents the number of links constructed by the 48 students for each pair of concepts.

Table 5.7 Number of links constructed by the 48 students for each pair of concepts (equations).

To / From	Equation	Linear equation in 1 unknown	Linear equation in 2 unknowns	Linear function	Unknown	Solution	$y = kx + b$	$ky + b = 0$	Proportional function	$ax + by + c = 0$	Total outgoing links from
Equation		39/38	35/35	7/5	22/22	36/34	2/2	2/1	3/2	0	146/139
Linear equation in 1 unknown	6/4		12/9	12/7	17/16	10/9	0	24/24	2/0	0	83/69
Linear equation in 2 unknowns	6/4	0		5/5	20/16	10/6	1/1	0	3/2	23/23	68/57
Linear function	7/1	9/9	5/5		5/5	0	24/23	1/0	21/18	1/1	73/62
Unknown	19/19	4/3	5/3	2/2		11/10	1/1	1/0	0	0	43/38
Solution	4/4	2/0	2/2	0	7/7		0	0	0	0	15/13
$y = kx + b$	0	2/1	2/2	14/13	2/2	1/0		13/13	3/2	1/1	38/34
$ky + b = 0$	1/0	10/9	1/1	1/1	5/5	4/3	8/6		1/1	1/1	32/27
Proportional function	4/4	2/0	3/3	17/15	3/2	3/1	3/1	0		1/1	33/26
$ax + by + c = 0$	0	0	9/9	0	2/2	3/1	2/2	0	0		16/14
Total incoming links to	47/36	68/60	74/69	58/48	83/77	75/63	41/36	41/38	33/25	27/27	547/479

Source: Jin and Wong, 2015, p. 696

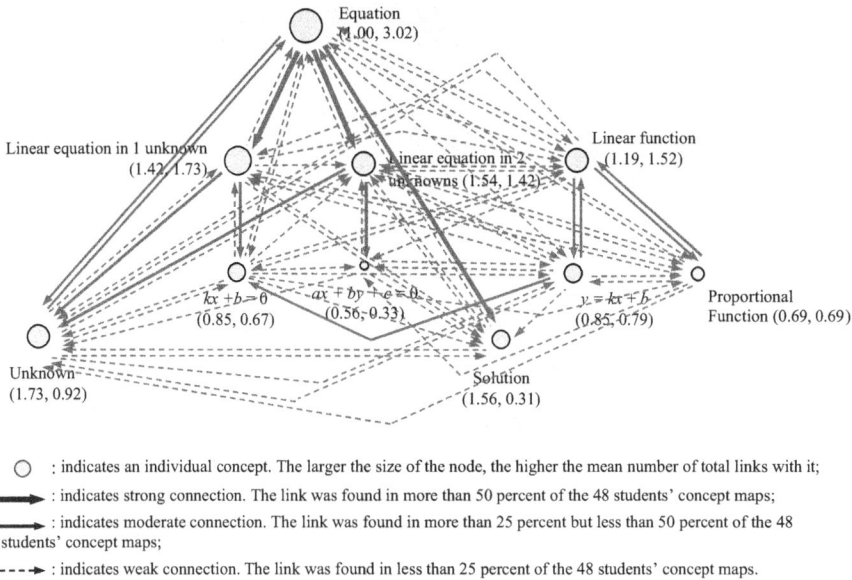

Figure 5.6 Collective map showing the connections between pairs of equation concepts.

Source: Jin and Wong, 2015, p. 697

The collective map in Figure 5.6 was constructed with reference to the number of links constructed among the given concepts, irrespective of whether the links were labelled with correct, partially correct, or incorrect linking phrases. Twelve and 24 are the cut-off points for the strong and moderate connections, respectively, in the collective map. The relative sizes of the nodes in the collective map are roughly based on the mean numbers of total links with the concepts.

As with algebraic expressions, the concepts are arranged according to the whole-part or set-subset relationships of the concepts, as indicated by their definitions in the students' textbooks. Equation is the most inclusive concept in the given concept list because the linear equations, functions, and their numerical representations can be seen as special types of equations. They are in a set-subset relationship with equation. Moreover, solution and unknown can be seen as parts of equation, because each equation includes unknown(s), and solution refers to the set of values that make an equation true. They are in a whole-part relationship with equation. Proportional function is a subset of linear function; it is positioned below linear function on the collective map. The three expressions, i.e. $y = kx + b$, $kx + b = 0$, and $ax + by + c = 0$ are numerical representations; they are positioned just below linear function, linear equation in one unknown, and linear equation in two unknowns, respectively.

As with the algebraic expressions mapping task, students appeared to be having difficulty building connections among the given concepts for this topic. They constructed fewer links compared with the previous other topic. Few students linked equations and functions. I offered the following prompts to encourage them to consider other possible connections.

1   What are the general expressions of linear equation in one unknown and linear equation in two unknowns? If the expressions are already given in the list, please connect them accordingly.
2   What is the general expression of proportional function? Does it relate to any of the concepts or expressions in the given list?
3   Is there any relationship *between linear equation in one unknown and linear function, or in general, between equations and functions*?
4   What is the connection between $kx + b = 0$ and $y = kx + b$? How about their graphs?

Considering only the moderate and strong links, the collective map indicates that the students generally divided the concepts into two clusters, one for equations and one for functions. Except for the moderate connection from $y = kx + b$ to $kx + b = 0$, all the links between the concepts in the two clusters are dotted ones. Even with the prompts, most students did not capture the relationship between equation and linear function in their maps. This is reasonable because equation and linear function were taught separately and no connection is explicitly presented in the textbook. Furthermore, in the Chinese mathematics textbooks, x in equations is called *unknown* (未知数), and x in functions is called *variable* (变量). This may account for students' limited understanding of the connections between the two concepts. Only 14 students built such connection, either from equation to linear function or from linear function to equation. Most such links reflected only partially correct or incorrect understanding of the relationships between the two concepts, e.g. 'is different from' and 'includes'.

Three pairs of concepts present moderate link strength in both directions, equation and unknown, linear function, and $y = kx + b$, and linear and proportional functions. This finding suggests that a substantial number of students found it roughly equal to link from one concept to another or vice versa for those pairs. The link from equation to unknown follows a whole-part sequence. They may have linked unknown to equation because of the order of the concepts in the definition of equation. They tended to link the concepts using the definition 'an equality which includes unknown(s) is called an (equation)' as the linking phrase.

For the links *between linear function and proportional function, most students correctly indicated in the linking phrases that the general expression of proportional function* is $y = kx$, in which, compared with the general *expression of linear function*, the constant term b equals zero.

The links *between linear equation in one unknown* and $kx + b = 0$, linear equation in two unknowns and $ax + by + c = 0$, and linear function and $y = kx + b$ were similar. These are relationships between the word representation of a concept and its numerical representation. For each of the three pairs, moderate links were constructed from the concepts to their corresponding expressions. However, only $y = kx + b$ was connected to linear function by a moderately strong link; the other two expressions were weakly connected to their corresponding word representations (as represented by dotted line links on the collective map). It may be that because students had newly learned linear function, they could easily retrieve information about it from memory.

The three strong links in the collective map indicate that the majority of the students were familiar with the links from equation to linear equation in one unknown, linear equation in two unknowns, and solution. The two special equations had moderate links with unknown and their numerical representations. The collective map illustrates that more students constructed a link from linear equation in two unknowns to unknown than from linear equation in one unknown to unknown.

The link from $y = kx + b$ to $kx + b = 0$ serves as the bridge connecting the function cluster and the equation cluster in the collective map. Only 27 percent of the 48 students indicated understanding of this connection in their concept maps, despite the prompts. Most students indicated that the two expressions have one same side or, when the value of y equals 0, the linear function can be seen as a linear equation in one unknown. Such connections are not quite insightful. Teachers may need to specify the relationships between equations and functions. Building the connection between functions and equations is meaningful. Secondary school students usually find functions abstract and difficult to learn. Connecting functions with equations can facilitate students' conceptual understanding of functions.

In sum, ideas about the relationships among equation concepts may differ among students. There are more weak links (dotted lines) in the collective map than moderate and strong ones. Over half of the students correctly connected the numerical representations with their corresponding word representations, but no student provided graphical representations of the concepts. The collective map captures gaps in the students' knowledge structure; such gaps must be bridged with appropriate lessons that, in this case, should include discussion on links across concepts in addition to tasks for mastering standard algebraic manipulation skills.

Concept Mapping on Triangles

Eleven triangle-related concepts were selected by the researcher from the students' textbooks for mapping. They were provided to the students in this order (translated from Chinese): triangle, acute-angled triangle, right-angled triangle, obtuse-angled triangle, scalene triangle, isosceles triangle, equilateral triangle, axis of symmetry, angle, median, and midline.

Individual Students Concept Maps (Triangles)

As in the previous sections, I present and describe a well-constructed concept map and a poorly constructed concept map to analyse the concept map's effectiveness

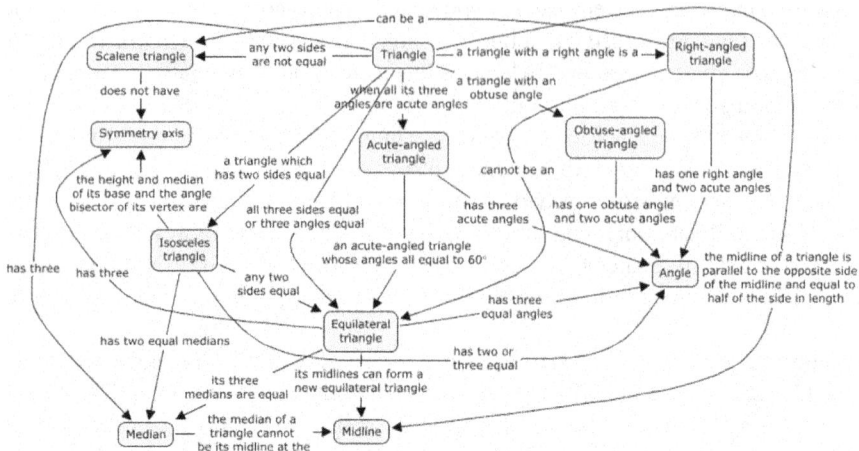

Figure 5.7 An example of a well-constructed concept map for triangle concept.

Note: Translated from Chinese

as a tool for assessing conceptual understanding in mathematics at the individual student level.

The well-constructed concept map presented in Figure 5.7 suggests that the student held a comprehensive understanding of the given triangle concepts.

All 11 given concepts are mapped, and no extra concepts were added. The concepts are not arranged in a top-down hierarchy but in more of a spider map format. The most inclusive concept triangle is at the centre, with the special types of triangles around it. The links from triangle to the six special triangles are labelled with definition-based linking phrases. The relationship between acute-angled triangle and equilateral triangle was built. The student pointed out that an acute-angled triangle whose angles are all equal to 60° is an equilateral triangle. However, the relationships between the other triangles are not reflected in the concept map, even though they were hinted at in the prompts. The relationship between an acute-angled triangle and equilateral triangle may have been more obvious to the student than the other relationships. Angle is in close proximity to acute-angled triangle, right-angled triangle, and obtuse-angled triangle, perhaps because triangles are often categorised according to their angles. Isosceles triangle and equilateral triangle both have the special properties of median and midline. Perhaps this is why the student positioned median and midline just below isosceles triangle and equilateral triangle. Axis of symmetry was placed near isosceles triangle. This is reasonable because, among the special types of triangles, isosceles triangle—other than equilateral triangle, a special type of isosceles triangle—is certainly the only type that has one axis of symmetry.

The student constructed 25 links for the 11 given concepts, presenting a density of $25/11 = 2.27$, indicating that, on average, each concept is linked more than twice. The density is higher than for most of the other students' concept maps on triangles. The propositions in the concept map are all correct, and some even include insights into the relationships. As each proposition scored 2 points, the map earned a proposition score of $2 \times 25 = 50$.

A poorly constructed concept map for the triangle concepts is presented in Figure 5.8. Although it did include all 11 given concepts, fewer links were constructed compared to the well-constructed map presented in Figure 5.7. Triangle is at the centre of the map, with seven concepts positioned around it. The links between triangle and the seven concepts are all labelled with definition-based linking phrases. No connections among the seven concepts are mapped. Equilateral triangle was positioned below isosceles triangle because, as presumably recognized by the student, equilateral triangle is an isosceles triangle with a 60° angle. Median and axis of symmetry might have been the last two concepts added to the map. On the one hand, there seems to be no space in the student's map for the two concepts to connect directly with triangle. On the other hand, the relationships between median and axis of symmetry with equilateral triangle seems to have been the most obvious one to the student. He simply linked the two concepts with equilateral triangle. He might not have attempted further connections

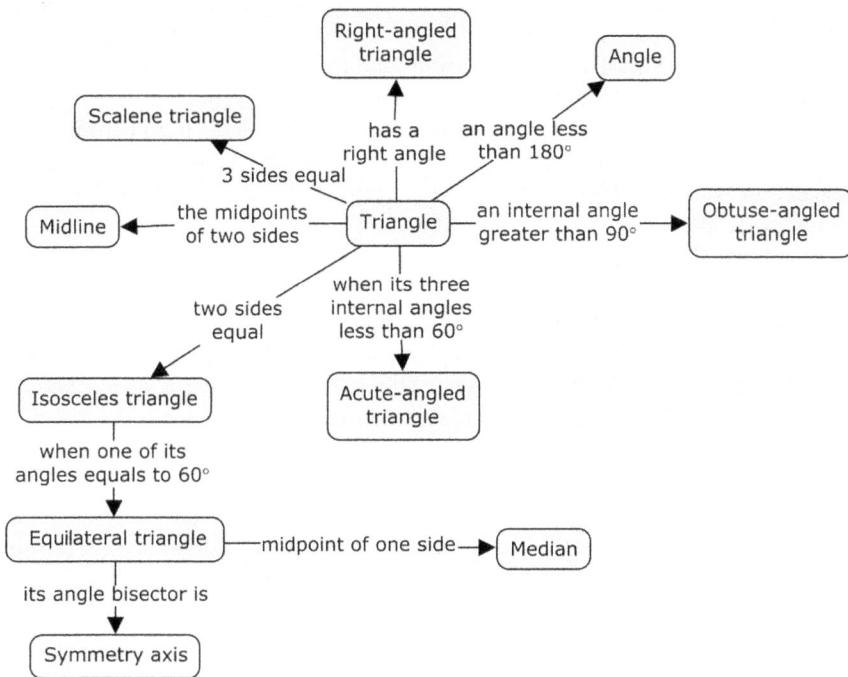

Figure 5.8 An example of a poorly constructed concept map for triangle concepts.

Note: Translated from Chinese

Table 5.8 Descriptive statistics of density and proposition score of concept maps and their correlation (triangles).

	Minimum	Maximum	Mean	SD	Spearman's Correlation
CM density	0.18	02.45	01.62	00.47	.821**
CM proposition score	0.00	48.00	25.75	11.50	

** Correlation is significant at the .01 level (2-tailed)

between median and axis of symmetry and the other concepts; that no concepts were left isolated could indicate that he was satisfied with the map.

The concept map presented in Figure 5.8 is a low-density map, 10/11 = 0.91, which is lower than 1.62, the mean density. This concept map earned a proposition score of $0+2+1+1+0+2+1+2+1+2=12$. This is far less than 25.75, the mean proposition score. The linking phrases indicate that the student may have had some understanding of the relationships but did not express that understanding clearly. For example, he linked triangle to midline with the linking phrase that 'the midpoints of two sides'. He may have known that the midline is a segment connecting the midpoints of the two sides of a triangle, but he did not include the information in the linking phrase.

The descriptive statistics of the densities and the proposition scores of the students' concept maps and their correlations are reported in Table 5.8.

The density of the concept maps of triangle concepts ranges from 0.18 to 2.45. A density of 0.18 indicates that only two links were constructed for the 11 given concepts, and at least seven concepts were left isolated. A density of 2.45 indicates that as many as 27 valid links exist among the given concepts, two more than the 25 links presented in the well-constructed map (Figure 5.7). Students proposition scores range from 0 to 48. The students who earned the highest and lowest proposition scores are those whose maps are the most and least dense, respectively. The Spearman's correlation of 0.821 ($p < .001$) indicates that the densities and the proposition scores are highly correlated. This is expected because concepts with higher densities offer more opportunities to construct propositions.

Links Associated With Individual Concepts (Triangles)

As in the previous sections, I analysed the concepts map's feasibility as a tool for assessing students' conceptual understanding in mathematics by individual concepts and by relationships between concepts in a pair at the whole-class level.

Individual Concepts (Triangles)

Table 5.9 displays the mean numbers of links connected to each given concept for the 48 students. The percentages of acceptable links with the concepts are

Table 5.9 Mean number of links of individual concepts (triangles).

Concept	Mean number of incoming links (I)	Mean number of outgoing links (O)	Mean number of total links (I + O)	Percentage of acceptable links
Triangle	0.08	6.98	7.06	79%
Isosceles triangle	2.19	2.04	4.23	73%
Equilateral triangle	1.48	2.73	4.21	73%
Right-angled triangle	1.35	1.92	3.27	72%
Acute-angled triangle	1.81	1.27	3.08	64%
Angle	2.81	0.27	3.08	85%
Scalene triangle	1.65	1.00	2.65	66%
Obtuse-angled triangle	1.27	1.10	2.37	71%
Symmetry axis	2.23	0.10	2.33	84%
Median	1.48	0.21	1.69	65%
Midline	1.44	0.17	1.61	77%

reported in the last column. The concepts are presented in descending order of the mean number of total links for ease of illustration.

Topping the list is the concept triangle. It had the highest number of total links to the other concepts, and almost all its links are outgoing. This is understandable because triangle is the most inclusive concept in the given list, and the other concepts are either its subsets, i.e. the six special types of triangles, or parts, i.e. axis of symmetry, angle, median, and midline. Another possible reason, which may also be a limitation of this study, is that triangle is the first concept presented in the given list. Students may begin mapping from the first given concept and build links from it. In this study, for each of the four concepts mapped, the most inclusive concept is also the first concept on the concept list; and so it is not possible to determine whether or to what degree the order of the given concepts affects the number of incoming and outgoing links of the concepts. This should be investigated in future studies.

As reported in the mean numbers of total links column in Table 5.9, among the six types of triangles, isosceles triangle features the most total links (4.23) with other concepts, followed by equilateral triangle (4.21). Both feature more outgoing links than the other types of triangles. Students may have been more familiar with isosceles triangle and equilateral triangle than with the other four types of triangles and thus been more able to readily conceive of their outgoing connections. The mean numbers of the total links to right-angled triangle and acute-angled triangle are 3.27 and 3.08, respectively; this is higher than the mean numbers of total links to scalene triangle (2.65) and obtuse-angled triangle (2.37). Isosceles triangle, equilateral triangle, and right-angled triangle may have been linked to most frequently because these three types of triangles have more properties than the other types of triangles, i.e. acute-angled triangle, obtuse-angled triangle, and scalene triangle. Students had more practice on such properties and had to have mastered them to solve problems. Accordingly, they were familiar with connections between isosceles triangle, equilateral triangle, and right-angled triangle, and the property-related concepts.

The percentages of acceptable links with the special types of triangles further support the claim that these students had conceptual knowledge of isosceles triangle, equilateral triangle, and right-angled triangle. These three types of triangles have higher percentages of acceptable links than the other three types. Acute-angled triangle has the lowest percentage of acceptable links. This is possible because some students defined acute-angled triangle in a similar manner as they did right-angled triangle and obtuse-angled triangle. Right-angled triangle is defined as a triangle with a right angle, and obtuse-angled triangle is a triangle with an obtuse angle. Some students mistook acute-angled triangle as a triangle with an acute angle, which is incorrect because every triangle has at least two acute angles. Acute-angled triangle is a triangle with three acute angles. The inaccurate definition of acute-angled triangle may have result in the partially correct propositions. For example, a proposition stating that acute-angled triangle has one acute angle is considered partially correct.

According to total number of links to the four property-related concepts, i.e. angle, axis of symmetry, median, and midline, students were able to conceive of more connections related to the angle properties of the triangles than to the median, symmetry, and midline properties. Most of the links connected to angle are incoming. This is possibly because angle is a part of all triangles. Students especially tended to build the links from triangles which are categorised according to their angle properties, i.e. acute-angled triangle, right-angled triangle, and obtuse-angled triangle. Moreover, isosceles triangle and equilateral triangle have particular properties related to angles. Isosceles triangle has at least two equal angles, and equilateral triangle has three equal angles. Students did not map as many connections to median, midline, and axis of symmetry as those to angles. Two possible reasons come to mind. First, mathematically, the triangles have few properties related to median, midline, and axis of symmetry. Not all triangles have an axis of symmetry; the property 'a midline of a triangle is parallel to the third side of the triangle and its length is one half of the length of the side' is shared by all types of triangles; only *right-angled triangle* has the property 'the median to the hypotenuse of a right-angled triangle equals to half the length of the hypotenuse' according to the textbook. The textbook includes few connections to *median, midline,* and *axis of symmetry.* Second, the students may not have been as familiar with the definitions of *median, axis of symmetry,* and *midline* as with other concepts, and were thus unable to connect them with related concepts.

Relationships Between Concepts in a Pair (Triangles)

This section reports the students' knowledge of the relationships between the given concepts according to their performance on the concept mapping task (triangle). Table 5.10 presents the number of links constructed by the 48 students for each pair of concepts.

Table 5.10 Number of links constructed by the 48 students for each pair of concepts (triangles).

To / From	Triangle	Isosceles triangle	Equilateral triangle	Right-angled triangle	Acute-angled triangle	Angle	Scalene triangle	Obtuse-angled triangle	Symmetry axis	Median	Midline	Total outgoing links from
Triangle		33/33	30/27	43/43	47/47	15/15	42/37	45/45	7/7	33/27	40/39	335/320
Isosceles triangle	0		18/17	6/6	3/2	21/21	1/0	2/2	34/32	10/10	3/3	98/93
Equilateral triangle	0	21/21		5/5	23/22	25/25	2/1	1/1	37/37	10/10	7/7	131/129
Right-angled triangle	0	27/27	5/4		2/1	23/23	21/19	1/0	3/1	7/7	3/3	92/85
Acute-angled triangle	0	12/10	11/11	5/1		23/23	3/3	2/0	1/1	1/1	3/3	61/53
Angle	2/2	1/1	0	0	0		0	0	9/9	1/1	0	13/13
Scalene triangle	0	1/0	3/2	5/5	11/11	4/4		9/9	12/12	1/1	3/3	49/47
Obtuse-angled triangle	0	10/9	2/2	0	1/0	24/24	10/7		1/1	2/2	3/3	53/48
Symmetry axis	0	1/1	2/2	0	0	0	0	0		1/1	1/0	5/4
Median	0	0	0	1/1	0	0	0	0	3/3		6/4	10/8
Midline	2/2	0	0	0	0	0	0	1/1	0	5/4		8/7
Total incoming links to	4/4	106/102	71/65	65/61	87/83	135/135	79/67	61/58	107/103	71/64	69/65	855/807

Source: Jin and Wong, 2021

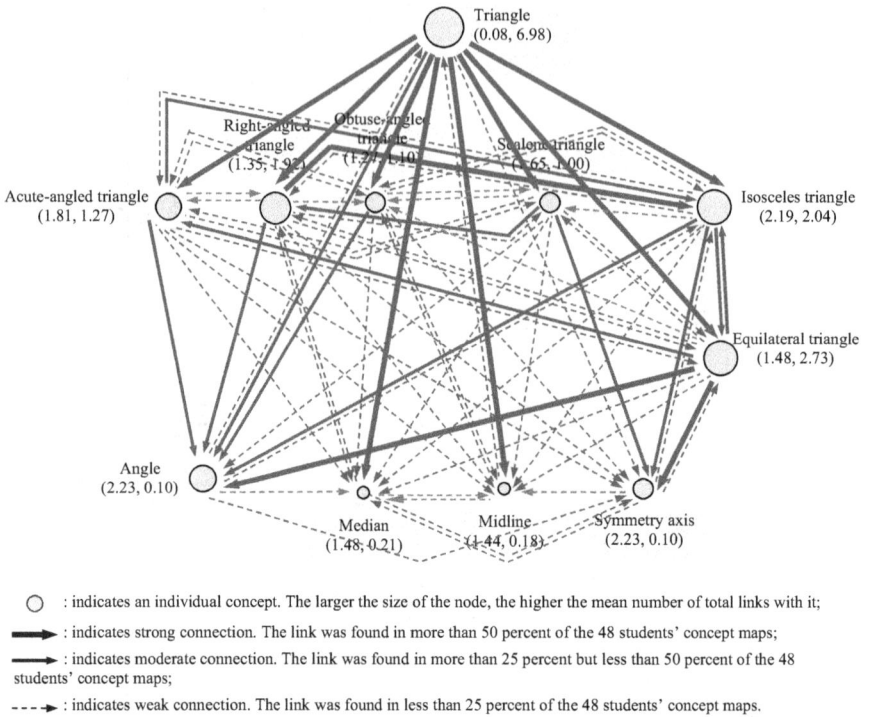

Figure 5.9 Collective map showing the connections between pairs of triangle concepts.

Source: Adapted from Jin and Wong, 2021

The corresponding collective map is presented in Figure 5.9. The hierarchies of the concepts are differentiated according to the superordinate, coordinate, and subordinate relationships indicated by the definitions in the textbook.

Triangle is the most inclusive concept among the concepts and is, accordingly, at the top of the collective map. Acute-angled triangle, right-angled triangle, obtuse-angled triangle, scalene triangle, and isosceles triangle, which are subsets of triangles, are placed below triangle. Since no set-subset or whole-part relationship exists between these triangles, they are at the same hierarchical level in the collective map. Equilateral triangle, which can be considered a subordinate concept of both isosceles triangle and acute-angled triangle, is lower in the hierarchy of the five types of triangles. The property-related concepts, i.e. median, axis of symmetry, angle, and midline, are at the bottom level of the collective map, because all the types of triangles possess some of these properties.

During the mapping session, it is observed that many students were not mapping connections among the coordinate (middle level) concepts. Prompts were issued to encourage them to think more deeply about the possible connections and how to express them so as to better demonstrate the breadth of their understanding of the concepts and their relationships. Most students did add connections to their maps after the following prompts:

1   Is there any relationship between *acute-angled triangle* and *equilateral triangle*? How about *right-angled triangle* and *equilateral triangle*?
2   Can a *right-angled triangle* be a *scalene triangle*? Can it be an *isosceles triangle*?
3   Can a *scalene triangle* be an *obtuse-angled triangle*? Can it be an *acute-angled triangle*?
4   What is the relationship between *midline* and *median* in triangles? Do they have special properties in special triangles?

In the collective map, most of the strong or moderate links are top-down, suggesting that students tended to link from higher hierarchical concepts to lower hierarchical concepts.

Triangle is the largest node because it has the most links with other concepts; isosceles triangle, equilateral triangle, acute-angled triangle, right-angled triangle, and angle are smaller nodes; obtuse-angled triangle, scalene triangle, and axis of symmetry are smaller still; the median and midline nodes are the smallest, because the mean numbers of total links with these two concepts are lower than the mean numbers of total links with the other concepts.

In general, the connections between triangle and most of the other concepts are very strong. They are labelled with definition-based linking phrases or linking phrases such as 'can be divided into' and 'includes'. By comparison, triangle and angle are only moderately related, and triangle and axis of symmetry are weakly related. This is possibly because triangle does not have the special properties of angles and axis of symmetry in general. Among the six types of triangles, equilateral triangle has the most properties related to angles and axis of symmetry. For example, it has three equal angles, three 60° angles, and three axes of symmetry. Accordingly, the connections between equilateral triangle and angle and between equilateral triangle and axis of symmetry are very strong. These strong connections suggest that most students understood the relationships.

Acute-angled triangle, right-angled triangle, and obtuse-angled triangle are categorised according to the properties of their angles' properties. The collective map illustrates that a moderate number of students mapped the connections and constructed the links between these triangles and angle. The other three types of triangles, scalene triangle, isosceles triangle, and equilateral triangle, have different axis of symmetry properties. The collective map reveals that a medium number of students built the connections between the triangles and axis of symmetry.

The first three prompts reminded students to consider the connections between different types of triangles. Twenty-three students (48 percent) linked equilateral triangle to acute-angled triangle, and 11 students linked from acute-angled triangle to equilateral triangle, suggesting that the majority of the students saw the two concepts as connected. That is, equilateral triangles are acute-angled triangles, and acute-angled triangles can be equilateral triangles if they have three equal angles or three 60° angles. Right-angled triangle and equilateral triangle have no substantial relationship with each other, and very few students linked them correctly with linking phrases such as 'has no connection with' or 'cannot be'. A triangle cannot be a right-angled triangle and an equilateral triangle at the same time. This might be why very few students linked the two concepts, as indicated by the dotted links in the collective map, despite the prompts to consider connections between types of triangles. More than 50 percent of the students mapped a link between right-angled triangle and isosceles triangle, and most of those students included a drawing of an isosceles right triangle to support their argument that a right-angle triangle can be an isosceles triangle and an isosceles triangle can be a right-angled triangle. This is likely because these students completed many exercises on isosceles right triangle and were required to flexibly apply the properties of isosceles triangle and right-angled triangle to solve assigned problems. This would account for the two concepts being well-connected in their collective knowledge structure. In contrast, only weaker (dotted) links were collectively mapped between obtuse-angled triangle and scalene triangle in the collective map because few students linked the two concepts. Students were likely familiar with obtuse-angled triangle and scalene triangle; the textbook offers few problems that address the properties of these two types of triangles, what to speak of the connections between them. This is likely why most students did not map meaningful connections between obtuse-angled triangle and scalene triangle and appeared to ignore any such connection.

As illustrated in the collective map, among all the pairs of concepts, students mapped bi-directional links (in this case moderately so) only between isosceles triangle and equilateral triangle, suggesting that students' collectively conceived that linking isosceles triangle to equilateral triangle or vice versa was acceptable.

Few students linked median and midline. Both concepts relate to the midpoints of the sides of a triangle. In isosceles triangle, the median to the base is perpendicular to the midline paralleling to the base. For equilateral triangles, the median length is $\sqrt{3}$ times the length of the midline. Accordingly, these two concepts can be connected in multiple ways. However, as suggested by the dotted links between them, most students were not aware of the relationships, perhaps because the two concepts were taught separately and the connection between them were never explicit presented to the students.

As illustrated in the collective map, few students built the link between axis of symmetry and median. If a triangle has an axis of symmetry, this axis of symmetry it is always the line which passes through one of its medians. Teachers may need to adjust instruction accordingly to ensure students are clear on those connections.

Analysis of the concept maps at the whole-class level indicates that the students accurately understood the given concepts and their relationships overall. They were able to build an appropriate number of connections among the triangle concepts. With the prompts, they were able to add connections among the coordinate concepts. In contrast to the mostly definition-based linking phrases we saw with the two algebraic topics, here most students included both definition- and property-based linking phrases. The collective map identifies the relationships students were most familiar with and those that required explication.

Concept Mapping on Quadrilaterals

Ten quadrilateral-related concepts were selected by the researcher from the students' textbook, ordered as follows (translated from Chinese): quadrilateral, rectangle, parallelogram, square, rhombus, trapezium, isosceles trapezium, diagonal, centre of symmetry, and axis of symmetry.

Individual Students Concept Map (Quadrilaterals)

As I have done earlier, I present a well-constructed concept map and a poorly constructed concept map to investigate the feasibility of the concept map as a tool for assessing conceptual understanding at the individual student level, this time with quadrilaterals.

The well-constructed concept map (Figure 5.10) includes the ten given concepts, which are hierarchically arranged in a top-down manner and connected through numerous links.

The most inclusive concept quadrilateral tops the map, followed by two major types of quadrilaterals trapezium and parallelogram. The student was apparently clear on the properties of the two types of quadrilaterals. She clearly stated that all parallelograms have centres of symmetry and that trapeziums, except for isosceles trapezium, do not have axes of symmetry. The relationships among the special types of quadrilaterals and between the quadrilaterals and the property-related concepts, i.e. axis of symmetry, centre of symmetry, and diagonal, are clearly

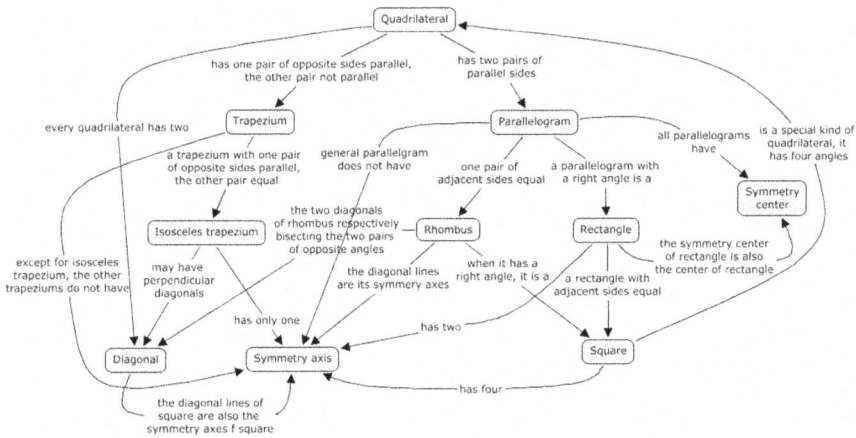

Figure 5.10 An example of a well-constructed concept map for quadrilateral concepts.

Note: Translated from Chinese

stated in the concept map. The student even considered the connection between diagonal and axis of symmetry, stating 'the diagonal lines of a square are also the axes of symmetry of the square'. Knowledge of that connection is useful when solving problems.

The student linked isosceles trapezium to diagonal with 'may have perpendicular diagonals', and she included a graph of an isosceles trapezium with perpendicular diagonals. Few students included graphic representations of the quadrilaterals in their maps. This graph, together with the linking phrase, is evidence of her comprehensive understanding of the diagonal property of isosceles trapezium.

She mapped 20 links among the ten given concepts. The density of the concept map is 20/10=2. All 20 links were labelled with accurate linking phrases, some indicating insights into the relationships, for a proposition score of 2 × 20 = 40.

The poorly constructed concept map (Figure 5.11) included only nine of the given concepts. *Centre of symmetry* is omitted, suggesting that the student was not familiar with it. The concepts are not hierarchically arranged. The mapping sequence is hard to follow, compared to the well-constructed concept map, Figure 5.10.

The student linked the most inclusive concept quadrilateral to rectangle, parallelogram, and square, possibly because she was most familiar with those three

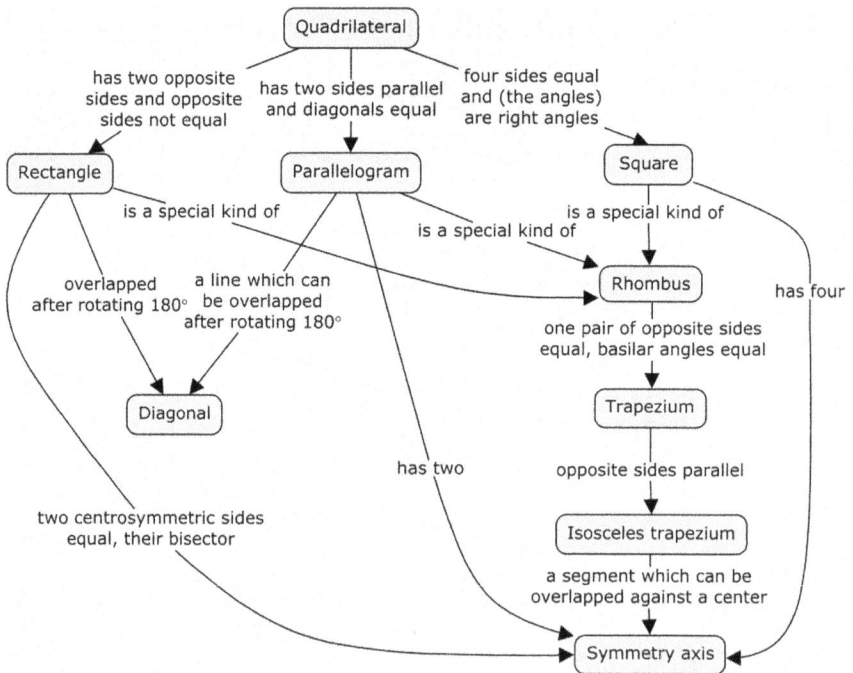

Figure 5.11 An example of a poorly constructed concept map for quadrilateral concepts.

Note: Translated from Chinese

Table 5.11 Descriptive statistics of density and proposition score of concept maps and their correlation (quadrilaterals).

	Minimum	Maximum	Mean	SD	Spearman's Correlation
CM density	0.10	03.00	01.75	00.54	.868**
CM proposition score	0.00	49.00	27.69	11.33	

** Correlation is significant at the .01 level (2-tailed)

concepts. These three types of quadrilaterals were linked to rhombus and labelled with the same linking phrase 'is a special type of'. Considering the directionality of links as introduced in the training, two of the propositions indicate an incorrect understanding of the relationships. Rectangle and parallelogram are not special types of rhombus; only the relationship between square and rhombus was correctly stated.

Many of the links in the student's concept map were labelled with incorrect linking phrases, suggesting limited understanding of the concepts and their relationships. Although the map is dense (14/10 = 1.4), the proposition score obtained is only $0+0+2+0+1+0+0+0+0+1+2+0+0+0=6$.

The descriptive statistics of the densities and the proposition scores of the students' concept maps and their correlations are reported in Table 5.11.

The density of the students' concept maps (quadrilaterals) range from 0.10 to 3.00 and as many as 30 links can potentially be built among the ten given concepts, although not all links are labelled with complete and correct linking phrases. Thirty is relatively high number of links because, mathematically, $(10 \times 9)/2 = 45$ is the maximum number of links that can be constructed for the ten given concepts. Students' proposition scores range from 0 to 49. The mean proposition score (27.69) is the highest among the four topics, indicating students were likely more familiar with the quadrilateral concepts than with the triangle and the algebraic concepts. The Spearman's correlation of 0.868 ($p < .001$) indicates that the two variables are highly correlated.

Links Associated With Individual Concepts (Quadrilaterals)

At the whole-class level, the students' concept maps are analysed for two aspects of conceptual understanding: the individual concept level and relationships between pairs of concepts level, and the findings are presented. The findings concerning these two aspects are described in subsequent sections.

Individual Concepts (Quadrilaterals)

The mean numbers of links connected to each given concept and the percentages of acceptable links is presented in Table 5.12. For ease of illustration, the concepts are presented in descending order of the mean number of total links.

Table 5.12 Mean number of links of individual concepts (quadrilaterals).

Concept	Mean number of incoming links (I)	Mean number of outgoing links (O)	Mean number of total links (I + O)	Percentage of acceptable links
Parallelogram	1.23	3.38	4.61	86%
Quadrilateral	0.29	4.21	4.50	86%
Square	2.30	1.98	4.28	80%
Rectangle	1.81	2.17	3.98	82%
Diagonal	3.17	0.73	3.90	78%
Rhombus	1.65	2.14	3.79	83%
Symmetry axis	3.39	0.14	3.53	78%
Isosceles trapezium	1.19	1.33	2.52	69%
Trapezium	1.02	1.23	2.25	75%
Symmetry centre	1.41	0.14	1.55	65%

Quadrilateral, the most inclusive concept among the given concepts has the highest mean number of outgoing links and the lowest mean number of incoming links among the given concepts. This is understandable because the other concepts in the given list are either subsets or parts of quadrilateral. The mean number of total links with quadrilateral is slightly lower than with parallelogram, perhaps because parallelograms have more properties than quadrilaterals. For example, a parallelogram has a centre of symmetry and a quadrilateral does not. The students can easily link parallelogram to centre of symmetry, while few possible links can be built between quadrilateral and centre of symmetry. The highest percentages of acceptable links were with quadrilateral and parallelogram, indicating that students knew well about the relationships between the two concepts.

Parallelogram and trapezium are differing branches of quadrilateral. Parallelogram has two pairs of parallel sides and trapezium has only one such pair. Collectively, students were more familiar with the parallelogram branch. They built more links to the concepts in this branch, i.e. parallelogram, rhombus, rectangle, and square. Moreover the percentages of acceptable links to this concepts is higher than for the trapezium branch concepts, i.e. trapezium and isosceles trapezium. Each concept in the parallelogram branch was linked with approximately four other concepts. By contrast, concepts in the trapezium branch were linked with approximately 2.5 other concepts.

Students mapped the least number of total links to centre of symmetry and of those, they mapped the lowest percentage of acceptable links among the given concepts. This is possibly because, on one hand, among the types of special quadrilaterals, only parallelogram has a centre of symmetry. This property is shared by the subsets of parallelograms, i.e. rectangle, rhombus, and square. After having linked parallelogram and centre of symmetry, students seldom followed-up with links with rectangle, rhombus, and square. On the other hand, students also tended to link parallelogram and centre of symmetry with the linking phrase 'has', which is partially correct according to the training parameters, and was awarded 1 point, resulting in a low percentage of acceptable links. An example of a proposition with a fully correct linking phrases indicating this connection is '*parallelogram* has a *centre of symmetry* which is the intersection of its two diagonals'.

The mean numbers of the incoming links with the three properties-based concepts, i.e. *diagonal, axis of symmetry*, and *centre of symmetry* are obviously higher than the corresponding mean numbers of outgoing links. The numbers of outgoing links and incoming links of these concepts suggest that students may tend to link more inclusive concepts to more specific and properties-based concepts.

Relationships Between Concepts in a Pairs (Quadrilaterals)

The number of links between concepts in a pair constructed by the 48 students is reported in Table 5.13. The concepts are arranged in the order of the given list. The numbers represent the number of students who built a link from one concept to another. Numbers higher than 12 are bolded.

The collective map (Figure 5.12) is based on the number of links between each concept in a pair (Table 5.13). The concepts are arranged in a hierarchical manner according to their superordinate, coordinate, and subordinate relationships, as indicated by their definitions in the textbook. Quadrilateral is superordinate to the other nine concepts and is positioned at the top of the collective map. Parallelogram and trapezium are subsets of quadrilateral and are accordingly positioned below quadrilateral. Similarly, rhombus and rectangle are below parallelogram and above square. Isosceles trapezium is below trapezium due to the set-subset relationship between them. The properties-related concepts, i.e. axis of symmetry, centre of symmetry, and diagonal, are at the bottom of the collective map because they can be seen as parts of quadrilaterals.

The collective map presents link strengths between paired concepts by line thickness and mean number of total links with a concept by node size. In this collective map, the strong and moderate links are all top-down, illustrating that students tended to construct links from the concepts higher in the hierarchy to those lower in the hierarchy. Students appear to have paid more attention to the hierarchies with quadrilaterals than with the other three topics and mapped more strong links, which may be related to the knowledge structure provided in the students' textbook (reproduced in Figure 5.13).

At the end of the chapter on quadrilaterals, the textbook summarises relationships among the key concepts presented therein, including the key concepts quadrilateral, parallelogram, trapezium, rectangle, rhombus, isosceles trapezium, right-angled trapezium, and square. The knowledge structure is a form of concept map, with concepts higher in the hierarchy positioned on the left and concepts lower in the hierarchy positioned on the right. The branches and hierarchies are clearly displayed. The relationships are specified. The graphical representations of the concepts are provided together with word representations. Similar knowledge structures were not included in the textbook for the other three topics.

Table 5.13 Number of links constructed by the 48 students for each pair of concepts (quadrilaterals).

To / From	Parallelogram	Quadrilateral	Square	Rectangle	Diagonal	Rhombus	Symmetry centre	Isosceles trapezium	Trapezium	Symmetry axis	Total outgoing links from
Parallelogram		46/46	16/16	36/35	19/15	38/36	28/28	1/1	4/4	14/9	160/148
Quadrilateral	4/4		27/24	36/33	19/18	25/25	3/2	7/7	42/40	2/2	207/197
Square	2/2	5/4		7/7	28/27	7/7	13/13	0	0	36/34	98/94
Rectangle	1/1	3/2	34/31		23/20	3/0	7/7	0	0	32/31	103/92
Diagonal	1/1	3/3	5/5	5/5		5/4	0	4/3		9/9	32/30
Rhombus	1/1	2/2	27/27	2/1	31/28		7/7	0	2/1	32/31	104/98
Symmetry centre	0	0	1/0	1/0	0	1/0		0	0	3/2	6/2
Isosceles trapezium	2/2	0	0	0	25/20	0	4/2		1/1	32/29	64/54
Trapezium	3/3	0	0	0	4/4	0	4/2	45/42		3/2	59/53
Symmetry axis	0	0	0	0	3/3	0	2/1	0	0		5/4
Total incoming links to	14/14	59/57	110/103	87/81	152/135	79/72	68/62	57/53	49/46	163/149	838/772

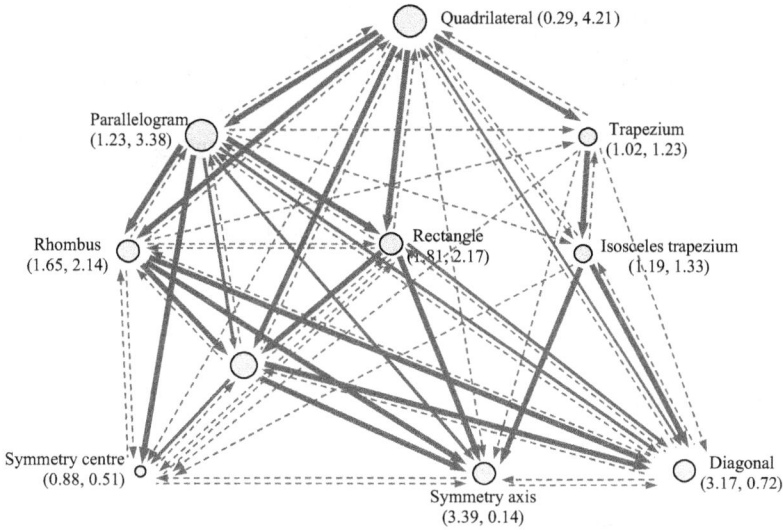

Figure 5.12 Collective map showing the connections between pairs of quadrilateral concepts.

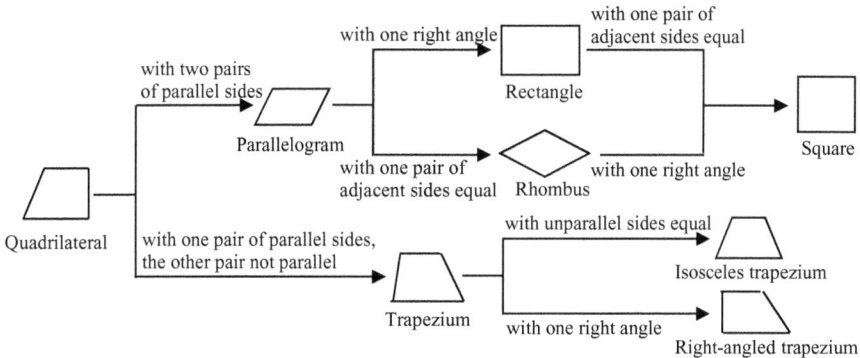

Figure 5.13 A knowledge structure in the textbook showing the hierarchical relationships of the quadrilaterals

Note: Translated from Chinese

Students' familiarity with the knowledge structure in the textbook likely enhanced their understanding and prepared them for concept mapping. This may be related to the greater abundance of connections and the more detailed and accurate linking phrases collectively presented. Comparatively fewer links were mapped from *quadrilaterals* to the properties-related concepts. To encourage the students to think about possible connections, the researcher gave the following prompts:

1   Does *parallelogram* have an *axis of symmetry*? Which types of *parallelograms* have *axes of symmetry*? How many do they have?
2   Can *isosceles trapezium* have perpendicular *diagonals*? If yes, can you give an example?
3   Do the *diagonals* of a *rhombus* have any special properties? How about the diagonals of the other types of quadrilaterals?
4   Under what conditions can the diagonal of a quadrilateral be its axis of symmetry?

In general, the connections illustrated in the knowledge structure (Figure 5.13) are all represented as strong links in the collective map. The concept right-angled trapezium was not given in the concept list for concept mapping. Five students included it in their concept maps, either as a concept or in a linking phrase. In addition to those connections represented in the knowledge structure, the collective map reflects strong links between quadrilateral and rhombus, rectangle, and square. Students labelled the connections between quadrilateral and these special types of quadrilaterals with definition-based linking phrases. Comparatively, fewer students directly linked quadrilateral to isosceles trapezium, but almost all students linked isosceles trapezium with trapezium. This may indicate that students find it easier to trace isosceles trapezium from trapezium than from quadrilateral, which is quite understandable because isosceles trapezium is defined based on trapezium.

The majority of the students linked parallelogram to centre of symmetry. Parallelogram has a centre of symmetry, which is the intersection of its two diagonals. This property is shared by all parallelograms. To avoid repeating the idea, students may have chosen not to link the special types of parallelograms with centres of symmetry. This might be why rhombus, rectangle, and square presented only moderate or weak links with centre of symmetry when analysed collectively.

The second prompt hinted at the relationships between parallelograms and axis of symmetry. Rhombus, rectangle, square, and isosceles trapezium are quadrilaterals with special properties of axis of symmetry. All presented strong connections with axis of symmetry in the collective map. Comparatively, trapezium was only weakly linked (dotted link) with axis of symmetry (because the general quadrilateral has no axis of symmetry property), and parallelogram is moderately linked with axis of symmetry. This is possibly because *parallelogram* is 'closer' to axis of symmetry than quadrilateral but further away than rhombus, rectangle, square, and isosceles trapezium. Most of the links from parallelogram to axis of symmetry are linked with 'may have', but five out of 14 such links were linked with 'has' or 'has one', indicating that students did not have a complete understanding of the properties of parallelogram. These findings suggest that students were familiar with the axis of symmetry properties of quadrilaterals. The link strengths may depend on whether a quadrilateral has an axis of symmetry property.

The third prompt was about the relationships between quadrilateral and diagonal. Quadrilateral is linked to diagonal with definition-based linking phrases or general linking phrase such as 'has' and 'has two'. Since trapezium has no specific diagonals property, the connection is presented as a dotted link on the collective map. Parallelogram, rhombus, rectangle, square, and isosceles trapezium are represented as either moderately or strongly connections with diagonal on the collective map, suggesting that students were quite familiar with the diagonal properties of the concepts.

The fourth prompt hinted at the possible relationship between diagonal and axis of symmetry; however, even with the prompt, only 11 students linked the two concepts, perhaps because they seldom explored the relationship during classroom exercises. Establishing the connection between these two properties-related concepts may help students navigate flexibly among different properties. Such flexibility is of value in solving problems.

Analysis of students' concept maps at the whole-class level indicates that students were familiar with quadrilateral concepts. They were able to construct numerous connections among the given concepts. Their familiarity with the knowledge structure of quadrilaterals from their textbook seemed to greatly impact their concept mapping. The hierarchies and connections reflected in the collective map are highly consistent with those in the knowledge structure. This finding suggests that such knowledge structures may be another valuable tool for assisting students in organising and structuring their knowledge.

Differences Among Topics

Differences were found between the students' concept maps for the algebraic topics and their maps for the geometric topics. They seemed more familiar with the geometric concepts than with the algebraic concepts. Compared with algebraic concepts, they generated relatively more links among geometric concepts and a higher percentages of the links constructed were acceptable. This finding should be considered in light of students' responses to a question I asked them in both a follow-up interview and as an open-ended written task (see Chapter 7): 'Do you find it easier to construct a concept map with algebraic concepts or with geometric concepts? Why?' Almost 90 percent of the 48 students answered that constructing a concept map with geometric concepts was easier. The two main answers were that geometric concepts are easy to remember and understand, while algebraic concepts are abstract and confusing and that more connections exist among geometric concepts than among algebraic concepts; other answered were that that geometric graphs help clarify the concepts and their relationships, that because they had learned algebraic concepts some time ago and had more recently learned geometric concepts, they were better able to recall the geometric concepts, and that algebra is mostly about calculation.

The collective maps on the two geometric topics include a greater number of strong and moderate links than the collective maps on the two algebraic topics. In particular, the collective map on equations has many weak links. The findings suggest that students' possessed similar knowledge structures for the two geometric topics and that they possessed markedly different knowledge structures for the two algebraic topics. This further suggests that psychologically, students' knowledge of the two geometric topics is more structured than their knowledge of the algebraic topics.

Most of the linking phrases in the concept maps on algebraic topics are definition-based, while the concept maps on geometric topics include primarily property-based linking phrases. This may be because geometric concepts are more concrete or object-like compared to algebraic concepts. Furthermore, geometric concepts are usually taught in terms of concrete properties, such as angles and lengths. Students use these properties to solve geometry problems. In so doing, they become familiar with properties and are eventually able to easily recall relevant properties during discussion. Algebraic concepts are comparatively more abstract, and students may require numerous examples and nonexamples to make sense of them. Moreover, school algebra usually involves manipulations and calculations. This may be why students thought of definitions, examples, and nonexamples when working with algebraic concepts.

Summary

Analysis of individual student's concept maps can provide meaningful information about their understanding of concepts, especially the relationships among the concepts, and address differences of students' understanding of given concepts. A missing connection may suggest that the student was not familiar with the connection or ran out of time to add the connection. If the missing connection is significant, teachers should intervene to assist the student in understanding the connection, if possible. Different students may provide different justification, even incorrect ones, for the same connection. Accordingly, investigating the linking phrases is important for assessing students' understanding of relationships and can provide teachers with detailed information about the extent to which their students have understood the concepts under study. On such basis, teachers can adjust their individual-level teaching accordingly. The densities and the proposition scores offer additional perspectives from which to evaluate students' conceptual understanding, as indicated by their concept maps.

Analyses of the concept maps at the whole-class level can indicate common strengths and weaknesses in students' conceptual understanding.

Concerning the individual concepts, the mean number of links reflects the degree to which a concept is involved with other concepts in a given domain, according to most students' understanding. The more links with a concept, the more important the concept may be within the domain. Relative connectedness of the given concepts is reflected in the size of the corresponding nodes. The percentage of acceptable links provides further information on students' common level of understanding of such links; the higher the percentage of acceptable links, the more thorough the students' understanding of connections among concepts in a topic.

The collective maps effectively capture distinct characteristics of students' conceptual understanding. The link strength (as indicated by line thickness) indicates how many students mapped the connection. A strong link (represented by the thicker line) usually indicates that most students were familiar with the connection. A common misunderstanding of an individual concept can result in commonly incorrect connections with the concept. Link direction of the strong links and moderate links (the thinner lines) in the collective maps suggest that the students tended to link from concepts of a higher hierarchy to concepts of a lower hierarchy. This finding suggests that students are familiar with the set-subset and whole-part relationships among mathematical concepts. However, they may not have the habit or ability to establish connections between coordinate concepts.

Note

The students' concept mapping on algebraic expressions and triangles has been published in the following two journal papers:

- Jin, H., and Wong, K. Y. (2015) 'Mapping conceptual understanding of algebraic concepts: an exploratory investigation involving Grade 8 Chinese students', *International Journal of Science and Mathematics Education*, 13(3), pp. 683–703.
- Jin, H., and Wong, K. Y. (2021) 'Complementary measures of conceptual understanding: a case about triangle concepts', *Mathematics Education Research Journal*, online-first version, doi.org/10.1007/s13394–021–00381–y

6 Comparing Concept Mapping and School Tests

Study Design

Comparative analysis is conducted on students' concept mapping performance and their performance on a definition-example-nonexample task (DEN task) and a paper-and-pencil task (P&P task) as more fully described in the following.

Participants

The same Grade 8 students who participated in the concept mapping training in Chapter 4 and the CM tasks described in Chapter 5 participated in the DEN and P&P tasks.

Instruments: DEN Tasks and P&P Tasks

I argue in Chapter 2 that definitions play an important role in the construction of mathematical concepts. Acquiring the definitions of individual concepts, along with examples and nonexamples is a first step toward conceptual understanding. The DEN tasks were designed to examine students' knowledge of definitions, examples, and nonexamples of the concepts in the given lists for the four mathematical topics used to test the concept map's feasibility as an assessment tool. The DEN task for *triangle* is presented as an example in Table 6.1.

To address the fact that most school mathematics tasks focus on students' problem-solving skills, I designed the P&P tasks specially to examine students' understanding of specific conceptual domains and, simultaneously, their ability to work with conceptual knowledge. Students completed four P&P tasks, one for each of the four topics featured in this book, i.e. algebraic expressions, equations, triangles, and quadrilaterals. Each P&P task includes items for assessing students' knowledge related to the three components of conceptual understanding (see Chapter 2) of the concepts in the given lists for the CM tasks (see Chapters 5). The items in the P&P tasks for the two geometric topics, triangles and quadrilaterals were adopted mainly from Usiskin's (1982) van Hiele geometry test and Hang's (1984) van Hiele-based geometry test. I designed the P&P tasks

DOI: 10.4324/9781003269373-6

Table 6.1 Definition-example-nonexample task (DEN task) (triangles).

No.	Concepts	Definition/ Description	Examples (Graph, Symbols, or Real-life Models)	Nonexamples
1	Triangle			
2	Acute-angled triangle			
3	Right-angled triangle			
4	Obtuse-angled triangle			
5	Scalene triangle			
6	Isosceles triangle			
7	Equilateral triangle			
8	Symmetry axis			
9	Angle			
10	Median			
11	Midline			

Source: Adapted from Jin and Wong, 2021

for the two algebraic topics, algebraic expressions and equations with reference to textbooks and reference books for secondary school mathematics in Singapore and China. The English items were translated into Chinese and the statements of the items were checked by the participants' mathematics teacher for clarity.

A pilot study was conducted on the P&P tasks with four Grade 8 classes in Nanjing, China. Each class participated in a P&P task for one of the topics. The tasks were then revised based on analysis of the data collected in the pilot study. Items with low discrimination were removed and new items were added. Two professors of mathematics education and the students' mathematics teachers examined the face validity of the P&P tasks. They agreed that the tasks addressed the students' understanding of the definitions, properties, relationships, and operations for the given concepts, rather than addressing procedural skills. The finalised P&P tasks and scoring rubrics are presented in Appendix D.

Data Collection

In the main study, the participants had 45 minutes to complete each P&P task during regular class time. They were assigned the corresponding DEN task and CM task (see Chapter 5) one to two days after. They had 15 minutes for each DEN task and 30 minutes for each CM task. The P&P tasks and DEN tasks were administered prior to the CM tasks because the participants had studied the four mathematical topics in class several months prior, and thus may have forgotten what they had learned. The P&P tasks and DEN tasks were intended to help them recall what they knew and prepare them for the CM tasks. The tasks were administered in Chinese and translated into English for reporting. During the data collection sessions, their mathematics teacher was present to make sure that they took the tasks seriously.

Data Analysis

The participants' responses to the P&P tasks are scored according to the typical method used in Chinese schools (see the marking scheme in Appendix D) for such tasks. The participants' mathematics teachers assisted in designing the marking scheme. Each participant earned an overall score for each P&P task. This P&P score is used in the correlational analysis with scores from the other two tasks.

The participants' answers to the definitions, examples, and nonexamples in the DEN tasks are coded and scored accordingly. For definitions, the students' responses are coded as correct, partially correct, incorrect, or no response. For examples and nonexamples, the answers are coded as correct, incorrect, or no response. Correct indicates accurate and detailed answers. Partially correct means that a reasonable answer was provided, but it is incomplete or only partially correct. Incorrect indicates an answer that is wrong or irrelevant. For example, consider the concept triangle. A correct definition is 'a closed plane figure formed by three line segments' or 'a figure formed by three line segments which are joined or can be joined end to end'. A definition like 'a figure formed by three line segments' is considered partially correct because it omits a key property, which is described by stating that the figure is closed or that the three segments are joined at their ends. Each correct answer is worth two points, partially correct answers 1 point, and incorrect answers (including no response) zero points. Accordingly, students receive a definition, example, and nonexample score for each of the four DEN tasks, all such scores are summed to arrive at the DEN task score.

In this chapter, in the following four sections which correspond to the four mathematical topics, findings from the DEN tasks and the P&P tasks are reported, each followed by comparison with the findings from the corresponding CM tasks. The correlations among the three types of tasks are investigated by correlating the concept map scores (i.e. proposition score and density) with the scores from the DEN tasks and the P&P tasks.

Algebraic Expressions

The 48 participants' performance on individual concepts (algebraic expressions) in the DEN task, the P&P task, and the CM task are reported in Table 6.2. For ease of comparison, though reported in Table 5.3 in the previous chapter, the mean numbers of total links with the concepts and the percentages of acceptable links of the concepts as assessed by the CM task (algebraic expressions) are repeated in Table 6.2. The concepts are sorted in ascending order by DEN score.

Algebraic Expressions: Definition-Example-Nonexample Task

The DEN task (algebraic expressions) involves the definitions, examples, and nonexamples of the given individual concepts. The top figures in the corresponding cells in Table 6.2 refer to the frequencies, and the bottom figures in parentheses are the percentages of 48 students. Correct answers are worth two points,

Table 6.2 Descriptive data of students' performance on individual concepts in the CM task (algebraic expressions), the DEN task (algebraic expressions), and the P&P task (algebraic expressions).

Concept	Definition-example-nonexample Task (DEN task)																	Paper-and-Pencil Task (P&P task)		Concept mapping Task (CM task)	
	Definition						Example					Nonexample					DEN-score	Percent of acceptable responses	Percent of points earned	Relevant item(s)	Mean number of total links	Percent of acceptable links
	No response	Wrong	Partially correct	Correct	Mean	SD	No response	Wrong	Correct	Mean	SD	No response	Wrong	Correct	Mean	SD						
Degree	4 (8.3)	15 (31.3)	19 (39.6)	10 (20.8)	0.81	0.76	3 (6.2)	13 (27.1)	32 (66.7)	1.34	0.96	8 (16.7)	14 (29.2)	26 (54.2)	1.08	1.00	155	54%	73%	4	2.55	77%
Coefficient	4 (8.3)	14 (29.2)	5 (10.4)	25 (52.1)	1.15	0.95	4 (8.3)	10 (20.8)	34 (70.8)	1.42	0.92	9 (18.8)	11 (22.9)	28 (58.3)	1.16	1.00	179	62%	53%	3	2.11	63%
Fractional expression	1 (2.1)	11 (22.9)	31 (64.6)	5 (10.4)	0.85	0.58	2 (4.2)	12 (25.0)	34 (70.8)	1.42	0.92	6 (12.5)	3 (6.2)	39 (81.2)	1.62	0.78	187	65%	84%	1	1.39	38%
Common factor	6 (12.5)	6 (14.6)	9 (18.8)	26 (54.2)	1.27	0.87	4 (8.3)	7 (14.6)	37 (77.1)	1.54	0.86	8 (16.7)	9 (18.8)	31 (64.6)	1.30	0.96	197	68%	68%	6	1.48	70%
Unlike terms	4 (8.3)	12 (25.0)	19 (39.6)	13 (27.1)	0.94	0.78	4 (8.3)	2 (4.2)	42 (87.5)	1.76	0.66	6 (12.5)	4 (8.3)	38 (79.2)	1.58	0.82	205	71%	84%	7	1.78	60%
Like terms	3 (6.2)	12 (25.0)	5 (10.4)	28 (58.3)	1.27	0.93	3 (6.2)	5 (10.4)	40 (83.3)	1.66	0.76	6 (12.5)	4 (8.3)	38 (79.2)	1.58	0.82	217	75%	84%	7	2.55	67%
Integral expression	0	13 (27.1)	3 (6.2)	32 (66.7)	1.40	0.89	3 (6.2)	1 (2.1)	44 (91.7)	1.84	0.56	3 (6.2)	10 (20.8)	35 (72.9)	1.46	0.90	225	78%	71%	1, 9(2)	3.19	48%
Monomial	1 (2.1)	8 (16.6)	9 (18.8)	30 (62.4)	1.44	0.80	2 (4.2)	2 (4.2)	44 (91.7)	1.84	0.56	5 (10.4)	7 (14.6)	36 (75.0)	1.50	0.88	229	80%	83%	2	4.70	68%
Constant term	1 (2.1)	3 (6.2)	4 (8.3)	40 (83.3)	1.75	0.60	1 (2.1)	5 (10.4)	42 (87.5)	1.76	0.66	4 (8.3)	2 (4.2)	42 (87.5)	1.76	0.66	252	88%	55%	5, 9(3)	1.57	58%
Polynomial	1 (2.1)	4 (8.3)	2 (4.2)	41 (85.4)	1.75	0.64	1 (2.1)	1 (2.1)	46 (95.8)	1.92	0.40	5 (10.4)	1 (2.1)	42 (87.5)	1.76	0.66	260	90%	71%	2, 9(4)	5.49	67%

partially correct answers one point, and wrong answers or non-responses worth zero points. The means and standard deviations of the scores for the definitions, examples, and nonexamples are reported accordingly in the table. The DEN-score for a concept is the sum of all the points the students earned on the definition, example, and nonexample for a given concept. This score is intended to reflect the level of students' understanding of the concept, as assessed by the DEN task. The percentage of acceptable responses for each concept in the DEN task is also calculated for direct comparison with the percentage of the acceptable links of the concepts in the CM task. It is the proportion of the DEN-score against the full mark that the 48 students could earn for an individual concept in the DEN task. The full mark is the same for each concept, $(2 + 2 + 2) \times 48 = 288$.

Students presented weaker performance in defining the concepts than in providing examples and nonexamples (Table 6.2). These differences are visualized in Figure 6.1, wherein the mean scores for the definition, example, and nonexamples items are plotted, by concept.

In Figure 6.1, the definition bars of the concepts are consistently shorter than the corresponding example bars and nonexample bars. According to students' exercise books and school mathematics tests, they had done more exercises requiring differentiation of examples from nonexamples than in drafting definitions, which may be one explanation for their better performance in providing examples and nonexamples. Difficulty expressing ideas clearly in words is another possible factor. The heights of the definition bars vary considerably, from the lowest scores 0.81 for degree and 0.85 for fractional expression to 1.75 for polynomial.

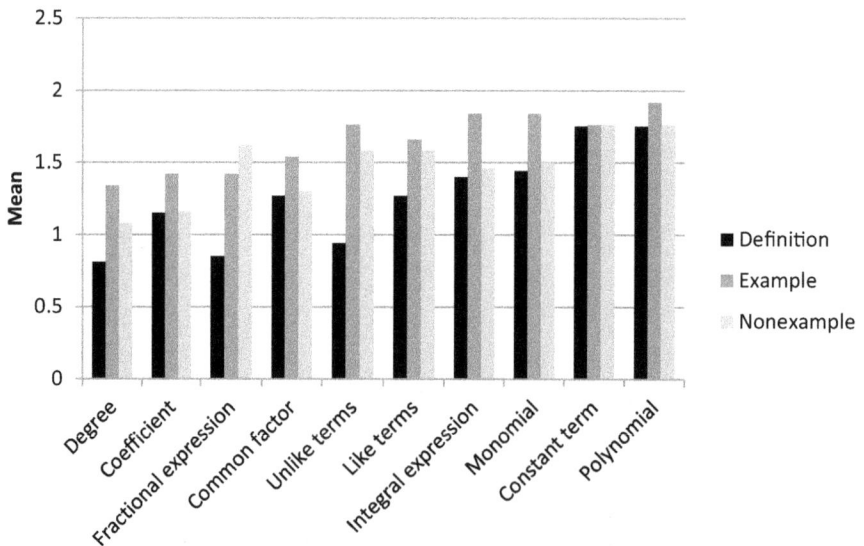

Figure 6.1 Means of definition, example, and nonexample scores of the algebraic expression concepts.

Students may have had more facility defining the more commonly used concepts, e.g. polynomial, compared to, e.g. fractional expression. Concepts such as degree are defined differently for monomials and polynomials. To obtain full points for the definitions, the students had to define the concept correctly for both cases; if they did not, the answer was scored partially correct. The height difference between the example and nonexample bars is less apparent. With the exception of fractional expression, students provided more correct examples than nonexamples. More students left the nonexample section blank. As they were seldom asked to provide nonexamples for the mathematical concepts, they were more familiar with examples.

The DEN-scores by concept suggest that the students had the most difficulty in understanding the concept degree. About 40 percent of the students provided partially correct definitions for *degree*. Some defined degree as 'exponents of variables in a monomial'; they did not specify 'the sum'. Other students defined degree as 'the sum of the exponents of variables in an expression'. Such a definition is inaccurate because it only applies to monomial, not to *polynomial*. The students were expected to provide examples of degree such as 'the *degree* of $a^2 + ab^2$ is 3' or simply '$a^2 + ab^2$, 3' such that others would get the idea that 3 is the *degree* of $a^2 + ab^2$. However, some 13 percent of the students presented examples like '$a^2 + ab^2$' or '$4a^2$' without specifying the degrees. With only the expression(s), the idea of *degree* cannot be captured; hence, such answers are considered incorrect. The vagueness of the examples and nonexamples may be another factor in the low DEN-score for degree. Similarly, many students gave vague examples and nonexamples of coefficient.

The students performed poorly on the DEN task for fractional expression. They had the most difficulty correctly defining this concept (see Table 6.2). Only five of 48 students correctly expressed that '*fractional expression* is an expression which can be written in the form of 'A/B' where A and B are integral expressions, B contains variable(s)'. More than half of the students (64 percent) provided partially correct answers. These answers can be classified as follows: (1) stating that fractional expression can be written in the form of 'A/B' and the denominator B contains variable(s) but not specifying that A and B are integral expressions (56 percent) and (2) stating that fractional expression can be written in the form of 'A/B' and that A and B are integral expressions but not specifying that B should contain variable(s) (6 percent). This finding suggests that students generally knew what a fractional expression was but could not describe it precisely in words. Almost 23 percent of the 48 students incorrectly defined fractional expression. They confused fractional expression with expressions that contain fractions, e.g. $2ab/3$. In addition, most of the wrong examples, e.g. $(a + b)/2$, A/B, and x^2/x, were provided by students who incorrectly or partially correctly defined fractional expression.

Some 25 percent of the 48 students appear to have misunderstood like terms and unlike terms. About 17 percent of the students thought that like terms have the same coefficient. Most made a similar mistake in defining unlike terms. They thought that unlike terms have different coefficients. Other students confused like terms and unlike terms with common factor and defined like terms as the common part of two terms, unlike terms as two terms which have no common parts. Partially correct definition scores for these two concepts are primarily due to incomplete answers. Other partially correct answers mentioned only special cases of unlike terms.

The DEN-scores show that, in general, the students had substantial knowledge of integral expression, constant terms, monomial, and polynomial. Even so, over 20 percent of the students wrongly defined *integral expression. Integral expression* is defined as follows in the textbook: 'monomial and polynomial together are called integral expression'. It is differentiated from *fractional expression* in whether or not the denominator contains variable(s). Some students defined *integral expression* as an expression which consists of numbers and variables. This is incorrect since it does not cover *polynomial* and, moreover, 'consist of' does not necessarily refer to multiplication. For example, $2/a$ consists of a number 2 and a variable a, but it is a fractional expression rather than a monomial or integral expression. Other students defined integral expression as an expression which does not involve fractions. This definition indicates that the students did not grasp the meaning of 'denominator does not contain variables' in the definition. A counter-example to the incorrect definition is $½ + a$, which involves the fraction $½$ but is an integral expression.

Algebraic Expressions: Paper-and-Pencil Task

The P&P task (algebraic expressions) includes items assessing students' understanding of individual concepts, relationships between concepts, and operations of the concepts and relationships. The marking scheme for the task is presented in Appendix D.

The percentages of points earned for the individual concepts are reported in Table 6.2, and are calculated by considering the mean points and the full points of the relevant items for the concept. For example, the P&P task examining students' knowledge of *integral expression* includes two items, Item 1 and Item 9(2). Students earned a full point score of 8 for Item 1 and a full point score of 1 for Item 9(2); they earned a mean points score of 5.368 for Item 1 and 0.729 Item 9(2). The percentage of points earned for *integral expression* is then calculated as $(5.368/8 + 0.729/1)/2 \times 100$ percent ≈ 71 percent. The same method is used to calculate the percentage of points earned for the other concepts. A higher percentage of points earned suggests better comprehension of the concept, as measured by the P&P task.

The percentages of points earned for the concepts suggest that the students may have had the most difficulty with *coefficient*. As indicated by Item 3, for expressions with concrete numbers as coefficients, e.g. $x^2 + 5x - 6$, students had no difficulty identifying the coefficients, but for expressions involving letters as constants, e.g. $-(2a)^2 x + (bx)^2 + ab^2x^2$, they could not distinguish constants from variables. For example, given that a is a constant, only 48 percent of the students answered correctly that the coefficient of x^2 in the expression $4ax^5 + 2x^3 - 7ax^2 + 2bx - a$ is $-7a$.

About 10.4 percent of the students answered $7a$, which is incorrect. They did not include the negative sign; they may have thought that the coefficient does not cover the positive or negative signs. Another 10.4 percent of the students answered '7' or '−7'. They seemed to mistake the coefficient of a term as the number in front of letters, regardless of whether the letter represents a variable or a constant.

Although the results of the DEN task indicate that the students had substantial knowledge of the definitions, examples, and nonexamples of constant term, the percentage of points earned in the P&P task indicates a weak understanding of this concept. For the relevant item, Item 5, students were to state the constant term(s) of two given expressions. Most students correctly stated the constant terms for expressions with letters representing the variables. However, for expressions with letters as constants, i.e. $a^2 - 2ab^2 + bx$, even though they were told that a and b were constants, over half of the 48 students did not recognise the constant term $a^2 - 2ab^2$. They may have thought that a constant term can only consist of number symbols. This misconception was also observed in the students' responses in Item 3.

The students' responses to the relevant item *degree* indicates that most students clearly recognised the degrees of the given expressions. For Item 4 of the P&P task, students were to state the degrees of four given expressions. They performed well with three of the expressions. However, for $5a - 8b + 2^2$, almost half the students mistook its degree as 2. They appeared to be looking for the highest exponent in the expression. This is not always correct because the degree of an expression relates only to the exponents of the variables. The exponents of the constants are not considered. For Item 10(2) students were to find the value of m such that the degree of $-0.7x^{(-0.5m - 4)} y^2 + x^2y - 3$ is 4, which requires correct application of the definition of *degree*. Over 75 percent of the students clearly stated that to have a degree of 4, it must be $-0.5m - 4 = 2$. These findings suggest that the students had a basic knowledge of *degree* but could not apply it well in non-routine cases.

The percentages of points earned for integral expression and fractional expression suggest that the students had a better understanding of fractional expression than of integral expression. This is opposite to the finding for the DEN task (algebraic expression). Item 1 relates to examples and nonexamples *of integral and fractional expressions*. It presents eight expressions; students were to identify whether they are integral expression, fractional expression, both, or neither. For expressions which are in 'standard' form, e.g. $a^2 + b + 7$ and a, nearly 90 percent of the students answered correctly that it is an *integral expression*. However, for expressions such as $(m - n)/5$, $2/5$, and $3ab + (2a/b)$, more than 30 percent of the students answered incorrectly. They tended to mistake expressions that expressed in fractional form or involving fraction(s) as *fractional expressions*. Item 9(2) asks directly whether 'an integral expression could include fraction(s)'. Over one quarter of the students (27 percent) thought that an *integral expression* cannot include a fraction. This finding is consistent with the inference made from their responses to Item 1. Although many students misunderstood the concept, about 90 percent of them still correctly solved the routine of *fractional expression* problem (Item 11). This raises an intriguing issue about the function of definition in problem solving, which should be investigated in future research.

Item 7 deals with examples and nonexamples of like terms and unlike terms. It includes four sub-items requiring students to decide whether the given terms are like terms or unlike *terms*. More than 90 percent of the students provided correct answers to three of the four sub-items, but only 40 percent of the students answered the fourth sub-item correctly. The majority of the students (60 percent) thought the two terms 4^3 and 5^2 are unlike terms, perhaps because the two numbers has different exponents. Whether two terms are like terms or unlike terms depends on the variables involved and the exponents of the variables. Item 10(3) also assesses students' understanding of the two concepts. It involves finding the value of the sum of m and n such that $2x^2y^m$ and $-x^ny^3/3$ are similar terms; this item requires direct application of the definition of like terms. More than 90 percent of the students answered correctly. These findings suggest that, although the students successfully recognised routine examples for like terms and unlike terms and performed well in solving routine problems, they did not fully understand the two concepts. Item 8 examines knowledge of the similarities and differences of like terms and unlike terms. Consistent with the findings for Items 7 and 10(3), the findings for this item also indicate that the students did not fully grasp the similarities and differences between the two concepts.

Algebraic Expressions: Correlations Among the Three Tasks

The prior two sections briefly summarise students' DEN task (algebraic expressions) and P&P task (algebraic expressions) performance. In this section, the relations between the three measures, the CM task (see Chapter 5), the DEN task, and the P&P task (algebraic expressions) are investigated by correlating the students' scores for the three tasks.

In the CM task (algebraic expressions), students used some of the corresponding concept definitions from the DEN task in the linking phrases, but few included the examples and nonexamples they presented in the DEN task. The DEN task (algebraic expression) directly addresses the examples and nonexamples of the given concepts; accordingly, it supplements some of the findings from the CM task. With regard to the individual concepts, findings from the two tasks regarding students' understanding are not always consistent. For example, students mapped the highest percentage of acceptable links (77 percent) with the concept *degree* in the CM task but provided the lowest percentage of acceptable responses (54 percent) for degree in the DEN task. With some concepts, consistent findings were observed. For example, performance on both the CM task and the DEN task indicate a weak understanding of fractional expression. Among the ten given concepts, students mapped the lowest percentage of acceptable links (38 percent) with fractional expression in the CM task and presented the third lowest percentage of acceptable responses (65 percent) in the DEN task. Such findings taken together suggest that the two tasks yield different conclusions about students' understanding of different concepts.

The student-constructed concept maps for algebraic expressions involved definitions of the individual concepts and relationships between the concepts, which are relevant to the first two components of conceptual understanding. The P&P task includes items relevant to all three components of conceptual understanding. It requires students to distinguish examples from nonexamples, recognise certain parts of given expressions, justify statements on a relationship, and apply knowledge of individual concepts in solving problems. Nevertheless, the P&P task is limited in that most of the included items are routine and address only isolated pieces of students' knowledge of the given concepts. Compared to the CM task, it is weak in assessing students' knowledge of the relationships between concepts. In addition, it scarcely explores students' justifications for their manipulations and operations. Such differences partially explain the inconsistencies between the P&P task (algebraic expressions) and the CM task (algebraic expressions).

Comparing the students' responses to Item 8 in the P&P task (algebraic expressions) and some of the propositions they constructed in the CM task (algebraic expressions) provides information which is useful in understanding inconsistent findings between the two tasks. Item 8 in the P&P task examines the similarity and difference between like terms and unlike terms and properties related to their coefficient and degree. It comprises four true-false statements (see Appendix D). Over 90 percent of the students knew that the statement 'like terms have same coefficients; unlike terms have different coefficients' and the statement 'like terms can be multiplied; unlike terms cannot be multiplied' are false. The percentages earned for the other two statements are relatively low, and students' misunderstanding of the relationships was revealed. Some 33 percent of the students indicated that the statement 'like terms have same degree; unlike terms have different degree' is true. Another 33 percent of the students considered the statement 'like terms can be added; unlike terms cannot be added' as false. The students' responses to the four statements lead to different assessments of their understandings of the concepts and their relationships. In the CM task (algebraic expressions), some students constructed incorrect propositions such as 'like terms have same coefficient', 'unlike terms have different coefficient', and 'unlike terms have different degrees'. However, a considerable number of students (over 50 percent) did not build the connections between like terms, unlike terms and coefficient and degree in their concept maps. If students do not provide the relevant propositions, it is unhelpful to compare their understanding of the concepts or relationships in the two tasks, because the absence of a connection in the concept maps does not necessarily mean that the student knew nothing about the relationship. This inference is supported by the students' observed reactions to the prompts (see Chapter 5). When prompted, most students added new propositions to their concept maps.

The relationships between the three tasks is further examined by correlating the scores the students earned for the tasks. For the CM task, each student earned a proposition score and a density score for their concept maps. For the DEN task, each student earned a definition score, an example score, and a nonexample score, which are the sum of the corresponding point(s) earned for the given concepts. For the P&P task, each student received an overall score, which is the sum of the points earned for all items in the task. Table 6.3 reports the results of the Spearman's correlation analyses.

Moderate to high levels of correlations (r ranged from .448 to .829, $p < .001$) are found among the scores. The proposition score has higher correlations with the DEN scores and the P&P score than with density, which suggests that compared to density, propositions may address a greater number of aspects of the students' conceptual understanding that are addressed by the DEN task and the P&P task. Factor analysis was conducted to investigate possible pattern(s) of the correlations. Given six task scores and 48 cases, this satisfies the minimum criterion for a sample size in that it comprises 'at least five times as many observations as there are variables to be analyzed' (Hair et al., 1998, p. 373). Table 6.4 reports the factor loadings of the task scores.

The result of the factor analysis indicates that the task scores yielded only one factor, with loadings all greater than 0.7. This supports the hypothesis that the three types of tasks measure the same underlying construct, *conceptual understanding*, despite different emphases among the tasks.

In summary, although the descriptive comparison indicates that the three types of tasks do not lead to consistent conclusions about students' understanding of the given concepts, both the factor analysis and the correlation analysis support the concurrent validity of the CM task as a technique for assessing students' conceptual understanding of algebraic expressions.

Table 6.3 Spearman's correlations between CM scores, DEN scores, and P&P scores (algebraic expressions).

		1	2	3	4	5
1	CM-density					
2	CM-proposition	.772**				
3	DEN-definition	.448**	.721**			
4	DEN-example	.572**	.733**	.702**		
5	DEN-nonexample	.545**	.633**	.595**	.829**	
6	P&P score	.685**	.739**	.740**	.732**	.698**

** Correlation is significant at the 0.001 level (2-tailed)

Table 6.4 Factor loading of the scores of the CM task, the DEN task, and the P&P task (algebraic expressions).

	Factor
	1
CM-proposition (algebraic expressions)	.901
CM-density (algebraic expressions)	.778
DEN-definition (algebraic expressions)	.876
DEN-example (algebraic expressions)	.887
DEN-nonexample (algebraic expressions)	.855
P&P score (algebraic expressions)	.884
% of variance	75%

Equations

Performance results for the 48 students on individual concepts in the CM task (equations), the DEN task (equations), and the traditional P&P task (equations) are reported in Table 6.5. The concepts are sorted in ascending order by the DEN-scores of the concepts. Due to the limitations in the design of the P&P task and the limited time available for testing, not all of the concepts in the given list were covered. For the concepts which have no corresponding item in the P&P task (equations), percentages of points earned are marked with '/' in the table.

Equations: Definition-Example-Nonexample Task

On the DEN task, most students provided correct examples and nonexamples for the given concepts. Their performance with definitions was relatively weak. The means of the definitions, examples, and nonexamples of the concepts are plotted in Figure 6.2. Except for the concepts $y = kx + b$ and *equation*, the bars representing definitions of the concepts are consistently shorter than the bars for examples and nonexample bars. For most of the concepts, the students seem to have been more proficient in providing examples than nonexamples.

As reported in Table 6.5, the students obtained the lowest DEN-score on linear equation in two unknowns. They had the greatest difficulty in defining this concept. A correct definition of linear equation in two unknowns should specify that it is an integral equation with two and only two unknowns and that the highest degree of the terms of the equation is one. Twenty-seven students (56 percent) presented partially correct definitions that stated only that the exponents of the unknowns are one. A counterexample of this definition is $xy + 1 = 0$. It is an integral equation with two unknowns x and y; however, it is not a linear equation because its highest degree is two, not one. Since 25 of the 27 students provided correct examples and nonexamples for this concept, students' inability to express their ideas clearly in words may have been a factor in the partially correct definitions. Among the ten wrong definitions, seven included only that linear equation with two unknowns is an equation with two unknowns. This definition is inaccurate because (*a*) it is not necessarily an integral equation, a typical counter-example being $x/y + 1 = 0$, and (*b*) it does not specify that the highest degree should be one, a counter-example of the definition being $x^2 + y = 0$. The other three incorrect definitions and corresponding examples suggest that the students confused linear equation in two unknowns with quadratic equation in one unknown.

Compared with their performance on linear equation in two unknowns, the students performed better in the definition and example sections of the DEN-score of linear equation in one unknown, indicating that they may have been more familiar with the latter. For the definition of linear equation in one unknown, as long as the students stated that it is an integral equation with one unknown and that the highest degree of the terms of the equation (or the exponent of the unknown) is one, the definition is considered fully correct. Even so, over 50 percent of the 48 students did not specify that it should be an integral equation. Without this condition,

Table 6.5 Descriptive data of students' performance on individual concepts in the CM task (equations), the DEN task (equations), and the P&P task (equations).

Concept	Definition-example-nonexample Task (DEN task)																		Paper-and-Pencil Task (P&P task)		Concept mapping Task (CM task)	
	Definition						Example					Nonexample					DEN-score	Percent of acceptable responses	Percent of points earned	Relevant item(s)	Mean number of total links	Percent of acceptable links
	No response	Wrong	Partially correct	Correct	Mean	SD	No response	Wrong	Correct	Mean	SD	No response	Wrong	Correct	Mean	SD						
Linear equation in 2 unknowns	2 (4.2)	10 (20.8)	27 (56.2)	9 (18.7)	0.94	0.67	3 (6.2)	8 (16.7)	37 (77.1)	1.54	0.84	4 (8.3)	0	44 (91.7)	1.84	0.56	207	72%	61%	1, 9(3)	2.96	62%
Unknown	0	13 (27.1)	17 (35.4)	18 (37.5)	1.10	0.81	2 (4.2)	2 (4.2)	44 (91.7)	1.84	0.56	5 (10.4)	8 (16.7)	35 (72.9)	1.46	0.90	211	73%	100%	7	2.65	69%
$ax + by + c = 0$	5 (10.4)	8 (16.7)	1 (2.1)	34 (70.8)	1.44	0.90	7 (14.6)	6 (12.5)	35 (72.9)	1.46	0.90	8 (16.7)	2 (4.2)	38 (79.2)	1.58	0.82	215	75%	/	/	0.89	89%
$kx + b = 0$	5 (10.4)	5 (10.4)	5 (10.4)	33 (68.8)	1.48	0.82	5 (10.4)	5 (10.4)	38 (79.2)	1.58	0.82	7 (14.6)	5 (10.4)	36 (75.0)	1.50	0.88	219	76%	/	/	1.52	74%
Solution	1 (2.1)	1 (2.1)	33 (68.7)	13 (27.1)	1.92	0.40	2 (4.2)	2 (4.2)	44 (91.7)	1.84	0.56	6 (12.5)	3 (6.2)	39 (81.3)	1.62	0.78	225	78%	76%	3, 9(5)	1.87	65%
Proportional function	1 (2.1)	8 (16.7)	8 (16.7)	31 (64.6)	1.46	0.80	2 (4.2)	9 (18.7)	37 (77.1)	1.54	0.84	3 (8.3)	3 (6.2)	41 (85.4)	1.70	0.72	226	78%	79%	2, 8(1,2)	1.38	60%
Linear function	2 (4.2)	9 (18.7)	7 (14.6)	30 (62.5)	1.40	0.84	2 (4.2)	3 (6.2)	43 (89.6)	1.80	0.62	5 (10.4)	3 (6.2)	40 (83.3)	1.66	0.76	233	81%	83%	2	2.71	74%
Linear equation in 1 unknown	2 (4.2)	5 (10.4)	26 (54.2)	15 (31.2)	1.17	0.66	2 (4.2)	1 (2.1)	45 (93.7)	1.88	0.48	4 (8.3)	0	44 (91.7)	1.84	0.56	234	81%	77%	1, 9(4)	3.15	72%
$y = kx + b$	2 (4.2)	3 (6.2)	5 (10.4)	38 (79.2)	1.69	0.66	3 (6.2)	2 (4.2)	43 (89.6)	1.80	0.62	5 (10.4)	6 (12.5)	37 (77.1)	1.54	0.84	241	84%	94%	9(6)	1.64	77%
Equation	0	3 (6.2)	5 (10.4)	40 (83.3)	1.77	0.56	1 (2.1)	1 (2.1)	46 (95.9)	1.92	0.40	3 (6.2)	5 (10.4)	40 (83.3)	1.66	0.76	257	89%	93%	9(1,2)	4.02	69%

/: no item in the P&P task addressing the students' knowledge of the individual concept

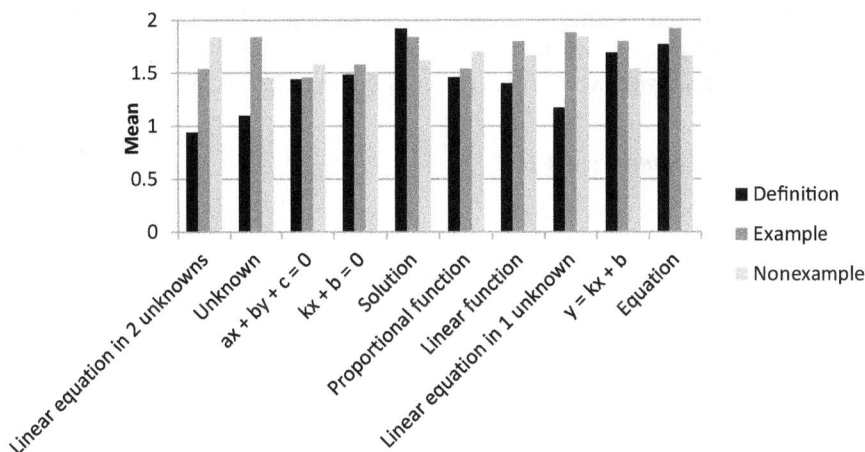

Figure 6.2 Means of definition, example, and nonexample scores of the equation concepts.

an equation is not necessarily a linear equation in one unknown. For example, $[2/(x+1)] + 1 = 0$ is an equation with only one unknown x. The exponent of x is one, but it is not a linear equation in one unknown.

Unlike the given concept lists for the other topics, the equation concepts given here involve both word and numerical representations. This could be a reason the majority of students defined linear function and proportional function via their numerical representations and correctly recognised that $y = kx + b$ is the general expression for linear function. About 63 percent of the 48 students defined linear function as a function which can be expressed in the form of $y = kx + b$ ($k \neq 0$). Among the seven partially correct answers, six did not include the specification that the coefficient k should not be zero. Among the nine incorrect definitions, three suggest the students confused linear function with proportional function, i.e. they stated that a linear function should be in the form of $y = kx$ which is only a special case of linear function. Other incorrect definitions include the suggestion that linear function is a function whose highest degree is one. Such a definition may have been adopted from the definition of linear equation in one (or two) unknown(s). The high rates of correct examples and nonexamples for linear function together suggest that the students were quite familiar with this concept, especially its numerical representation. Perhaps because they had a lot of practice with the numerical representation. With only one exception, the students who did not respond or who wrongly defined linear function, also defined proportional function incorrectly. Most of the definitions of proportional function received the designation partially correct, because the students did not specify that the coefficient k in the expression $y = kx$ should not be zero. Moreover, the students who provided incorrect examples or nonexamples of proportional function are the same students who provided incorrect or partially incorrect definitions. This suggests their provision of wrong or partially correct definitions of proportional function was not due to their inability to express ideas in words. These students did not understand what proportional function is and thus would likely have had difficulty solving exercises involving this concept until their understanding improved.

Equations: Paper-and-Pencil Task

The P&P task (equations) includes items addressing students' understanding of the given concepts (see Appendix D).

As indicated by the percentages of points earned (see Table 6.5), the students performed well on the items on unknown, $y = kx + b$, and equation. With a given statement and a corresponding equation, all students correctly recognised what the letter k represents in the equation. However, they seemed to possess only the basic knowledge of $y = kx + b$ and could not apply that knowledge effectively in solving problems. For example, for Item 6 students were to choose from four given graphs representing the function $y = kx - k$ ($k \neq 0$) such that the value of y decreases while the value of x increases. To choose correctly, students had to know the meaning of the coefficient k and the constant b in the general expression and clarify the relationship between the two. Many students (43.8 percent) answered the item incorrectly.

The percentage of points earned for the concepts shows that the students performed most poorly on the items about linear equation in two unknowns (percentage of points earned: 61 percent). The percentage of points earned for linear equation in one unknown (77 percent) is also relatively low. For Item 1, students were to identify examples and nonexamples of linear equations in one unknown and linear equation in two unknowns among 15 given expressions. These two concepts are firstly equations. All students clearly recognised that $x - 3y$ is neither a linear equation in one unknown nor a linear equation in two unknowns since it has no equal sign. Nearly 90 percent of the students correctly identified the 'standard' examples and nonexamples, i.e. $2x + y = 10$, $2x^2 - y = 9$, $7 + 3x = 14$, $3x - \frac{1}{4} = 8$, and $x/2 = 1$. However, some students had difficulty in recognising nonlinear equations in two unknown in which both unknowns have an exponent of one. About 30 percent of the students mistook $xy = 3$ as a l*inear equation in two unknowns*, and 42 percent of the students answered that $xy = x - y$ is a linear equation in two unknowns. For equations which involve fraction(s) or in fractional forms, e.g. $2x + 2/y = 1$, $x/4 + y/3 = 1$, and $(y - 1)/3 = 2y - 3$, more than 20 percent of the students did not correctly distinguish between linear equations in one unknown, linear equation in two unknowns, and 'neither'. In addition to these findings, the students' responses to the equation $s = 5t$ indicate that they may have been more familiar with equations wherein the letters x and y represent unknowns. Two students miscategorised $s = 5t$ as a linear equation in one unknown, and eight students (17 percent) responded that it was neither a linear equation in one unknown nor a linear equation in two unknowns. Twenty-seven percent of the 48 students mistook $x^2 - 2x - 3 = 0$ as a linear equation in two unknowns. This finding is consistent with the finding in the DEN task in that some students confused linear equation in two unknowns with quadratic equation in one unknown. Such misconception may be because, in Chinese, the terms used for these two concepts are similar: one is called 二元一次方程 and the other is called 一元二次方程. The students may not be clear that '元' refers to the number of unknowns while '次' refers to the highest degree of the equation.

Item 9(3) involves the definition of *linear equation in two unknowns*. The statement addresses a common mistake made by secondary students. The DEN task yielded a similar finding: 65 percent of the students mistook it as a correct statement and apparently thought that if an equation has two unknowns and the exponent of each unknown is one, the equation is a linear equation in two unknowns. Item 9(4) asks whether the coefficient of the unknown in a linear equation in one unknown should be one. This item was designed to address some students' misunderstanding of the definition; some students mistook the 'highest degree of the terms of the equation is one' as the 'exponent of unknowns is one'. Forty-two percent of the students answered this item incorrectly. This finding is consistent with the finding for the DEN task, wherein students had the most difficulty in correctly defining linear equation in two unknowns (see Table 6.5). Nevertheless, the students performed quite well on the items (i.e. Items 12 and 13) on the application of the concepts *linear equation in one unknown* and *linear equation in two unknowns*. The findings together show that although the majority of the students did not accurately define the two concepts and also had difficulty recognising 'non-standard' examples and nonexamples, they solved the relevant problems.

The percentages of points earned for proportional function and linear function are 79 percent and 83 percent, respectively. Item 2 presents examples and nonexamples of these two concepts. The students were to indicate whether a given equation was a linear equation, a proportional equation, both, or neither. They had the most difficulty with the following two equations: $y = -x/3$ and $y = 8x^2 + x (1-8x)$. Nine students indicated that $y = x/3$ was neither a linear function nor a proportional function; four students indicated that it was a linear function. The relatively poor performance may be because the coefficient of x in this equation is the negative number $-1/3$. As in the DEN task (equations), some students misconceived that the two variables of a proportional function must be in a proportional relation and required the coefficient to be positive. For the equation $y = 8x^2 + x (1-8x)$, the students had to first simplify. About 48 percent of them recognised that it is actually a proportional function. Fifteen percent of the students recognised it as a *linear function*, but 37.5 percent identified it as 'neither' a linear function nor a proportional function. The students who chose neither may not have considered the need to simplify the equation and mistook it as a quadratic equation.

Equations: Correlations Among the Three Tasks

This section examines the relationships among the CM task (equations), the DEN task (equations), and the P&P task (equations). The earlier analyses of the three tasks suggest that the tasks assess different aspects of students' understanding of the given concepts. The CM task (equations) focuses on knowledge of the relationships between the equations. Many of the connections were labelled with definition-based linking phrases. The concept maps reveal misconceptions like 'the coefficient of unknown in linear function in one unknown should be one' and also indicate that the students were unfamiliar with the connections between equations and functions. Such information is not revealed from the DEN task (equations) or the P&P task (equations). The DEN task involves definitions, examples, and nonexamples of individual concepts. The students showed more proficiency in providing examples than nonexamples of most of the concepts. This finding is not available from the CM task since few students presented examples or nonexamples of the concepts in their concept maps. The P&P task (equations) are more focused on application of the definitions and calculations. Few of the items in the task address relationships between the given concepts. However, the P&P task (equations) shows that the students had difficulty in recognising non-routine examples and nonexamples. Such a finding is not easily obtainable from the other two types of tasks.

Together, the three tasks lead to inconsistent conclusions about students' understanding of some concepts. For example, the results of the CM task (equations), as indicated by the percentages of acceptable links, suggest that the students had a good understanding of $ax + by + c = 0$. This expression has an acceptable link percentage of 89 percent, the highest among the ten given concepts. Results of the DEN task (equations), however, indicates that over a 25 percent of the students defined this concept incorrectly. The students' responses to Items 4 and 13 in the P&P task (equations) also suggest that they had some difficulty with $ax + by + c = 0$. The percentages of students who earned the full marks for the two items are 69 percent and 75 percent, respectively.

The relationships between the three tasks are examined by considering the Spearman's correlations of their task scores. The correlation coefficients are presented in Table 6.6.

The CM task scores (CM-density and CM-proposition) are significantly correlated with the DEN task scores (DEN-definition, DEN-example, and DEN-nonexample) and the P&P task score. The correlation between the proposition score and the DEN-definition score is the highest, $r = .734$, $p < 0.001$. This high correlation

Table 6.6 Spearman's correlations between CM scores, DEN scores, and P&P scores (equations).

		1	2	3	4	5
1	CM-density					
2	CM-proposition	.794**				
3	DEN-definition	.472**	.734**			
4	DEN-example	.382**	.564**	.762**		
5	DEN-nonexample	.513**	.582**	.744**	.803**	
6	P&P- score	.458**	.711**	.852**	.741**	.635**

** Correlation is significant at the 0.001 level (2-tailed)
* Correlation is significant at the 0.01 level (2-tailed)

Table 6.7 Factor loading of the scores of the CM task, the DEN task, and the P&P task (equations).

	Factor
	1
CM-proposition (equations)	.862
CM-density (equations)	.696
DEN-definition (equations)	.913
DEN-example (equations)	.849
DEN-nonexample (equations)	.848
P&P score (equations)	.879
Percentage of variance	71%

may be due to students' use of numerous definition-based linking phrases to connect the equation concepts. The correlation between the proposition score and the P&P score is also high and significant ($r = .711, p < .001$). Moreover, the P&P score is highly correlated with the DEN-definition score ($r = .852, p < .001$). The findings together suggest that students who did well in defining the equation concepts may tend toward good performance in the P&P task (equations). They might also tend to construct more connections among the given concepts and label them with meaningful linking phrases.

Factor analysis was conducted to investigate the correlation patterns. Table 6.7 reports the factor loadings of the task scores. All scores have high loads on a single factor which accounts for 71 percent of the variance. This finding suggests that the three tasks measure the same underlying variable. However, the three types of tasks measure different aspects of students' understanding of the given concepts; both the factor analysis and the correlation analyses suggest that the aspects are highly correlated. This finding is similar to findings for the topic *algebraic expressions*.

Triangles

Table 6.8 reports the results of the students' performance on the individual concepts in the CM task (triangles), the DEN task (triangles), and the P&P task (triangles). The concepts are sorted in ascending order of their DEN scores rather than in the order in which they are presented in the given list.

Triangles: Definition-Example-Nonexample Task

Table 6.8 indicates that the students performed well with the definitions, examples, and nonexamples of the six special types of triangles, i.e. isosceles triangle, obtuse-angled triangle, acute-angled triangle, scalene triangle, equilateral triangle, and right-angled triangle. An exception is with nonexamples of isosceles triangle. In the textbook, isosceles triangle is defined as a triangle with at least two

Table 6.8 Descriptive data of students' performance on individual concepts in the CM task (triangle), the DEN task (triangle), and the P&P task (triangle).

Concept	Definition–example–nonexample Task (DEN task)																	Paper-and-Pencil Task (P&P task)		Concept mapping Task (CM task)	
	Definition						Example					Nonexample					DEN-score	Percent of acceptable responses	Percent of points earned	Relevant item(s)	Mean number of total links	Percent of acceptable links
	No response	Wrong	Partially correct	Correct	Mean	SD	No response	Wrong	Correct	Mean	SD	No response	Wrong	Correct	Mean	SD						
Median	2 (4.2)	16 (33.3)	2 (4.2)	28 (58.3)	1.21	0.97	3 (6.2)	8 (16.7)	37 (77.1)	1.54	0.84	7 (14.6)	1 (2.1)	40 (83.3)	1.66	0.76	212	74%	/	/	1.69	65%
Symmetry axis	1 (2.1)	11 (22.9)	25 (52.1)	11 (22.9)	0.98	0.70	1 (2.1)	1 (2.1)	46 (95.8)	1.92	0.40	1 (2.1)	6 (12.5)	41 (85.4)	1.70	0.72	221	77%	/	/	2.33	84%
Angle	0	5 (10.4)	20 (41.7)	23 (47.9)	1.38	0.67	1 (2.1)	0	47 (97.9)	1.96	0.28	6 (12.5)	7 (14.6)	35 (72.9)	1.46	0.90	230	80%	/	/	3.08	85%
Midline	1 (2.1)	4 (8.3)	4 (8.3)	39 (81.2)	1.71	0.65	2 (4.2)	3 (6.2)	43 (89.6)	1.80	0.62	5 (10.4)	3 (6.2)	40 (83.3)	1.66	0.76	248	86%	/	/	1.61	77%
Triangle	0	4 (8.3)	27 (56.2)	17 (35.4)	1.27	0.61	0	0	48 (100)	2.00	0.00	0	0	48 (100)	2.00	0.00	253	88%	96%	1, 2	7.06	79%
Isosceles triangle	0	1 (2.1)	2 (4.2)	45 (93.7)	1.92	0.35	0	0	48 (100)	2.00	0.00	0	12 (25.0)	36 (75.0)	1.50	0.88	260	90%	84%	7	4.23	73%
Obtuse-angled triangle	0	6 (12.5)	0	42 (87.5)	1.76	0.66	0	4 (8.3)	44 (91.7)	1.84	0.56	1 (2.1)	2 (4.2)	45 (93.7)	1.88	0.48	262	91%	/	/	2.37	71%
Acute-angled triangle	0	8 (16.7)	0	40 (83.3)	1.67	0.75	0	2 (4.2)	46 (95.8)	1.92	0.40	0	2 (4.2)	46 (95.8)	1.92	0.40	264	92%	92%	4(1, 4)	3.08	64%
Scalene triangle	0	4 (8.3)	0	44 (91.7)	1.83	0.56	0	4 (8.3)	44 (91.7)	1.84	0.56	0	2 (4.2)	46 (95.8)	1.92	0.40	268	93%	66%	5(1, 3, 4)	2.65	66%
Equilateral triangle	0	2 (4.2)	0	46 (95.8)	1.92	0.40	1 (2.1)	0	47 (97.9)	1.96	0.28	1 (2.1)	1 (2.1)	46 (95.8)	1.92	0.40	278	97%	/	/	4.21	73%
Right-angled triangle	0	3 (6.2)	0	45 (93.8)	1.88	0.48	0	0	48 (100)	2.00	0.00	0	1 (2.1)	47 (97.9)	1.96	0.28	280	97%	/	/	3.27	72%

/: no item in the P&P task addressing the students' knowledge of the individual concept

Source: Adapted from Jin and Wong, 2021

equal sides. The textbook defines equilateral triangle as a special kind of isosceles triangle. However, nearly 25 percent of the 48 students used equilateral triangle as a nonexample of isosceles triangle. This finding indicates that these students did not quite understand that equilateral triangle is actually a subset of isosceles triangle.

They may have faced difficulty defining some geometric concepts accurately. All 48 students provided correct examples and nonexamples of *triangle*; however, the percentage of correct definitions of *triangle*, about 35 percent, is comparatively low. *Triangle* is defined as a closed plane figure formed by three line segments. Of the 27 partially correct definitions of *triangle*, 17 included that a triangle is a figure which consists of three segments but did not include that the three segments must be joined end to end.

Compared to their performance on the other concepts, the students' performed relatively poorly on definitions, examples, and nonexamples of the four property-related concepts, *i.e.* median, axis of symmetry, angle, and midline.

The students obtained the lowest DEN-score on *median*. Some 33 percent of the 48 students defined it incorrectly. These students normally knew that median relates to the midpoint of a segment or that it bisects a segment but could not express the idea clearly. For example, eight students defined it as a line which divides a segment into two equal parts, two students confused it with a perpendicular bisector, and three students defined it as a segment which divides a triangle into two (equal) parts. The eight incorrect examples were all provided by students who submitted incorrect definitions.

The students had the greatest difficulty in defining *axis of symmetry*; only 11 of the 48 students (23 percent) submitted a fully correct definition of *axis of symmetry*, and almost 50 percent of the 48 students provided only a partially correct definitions. One possible reason is simply that that the definition of *axis of symmetry* is more demanding than the other concepts. Axis of symmetry is defined as a line in a plane figure which divides the figure into two parts such that one part, when folded over along the axis, shall coincide with the other part. This definition requires several components: 'a line in a plane figure', 'divides the figure into two parts', and 'shall coincide with each other'. The students might find it hard to remember them all. Twenty students (42 percent) defined it as a line in a plane figure which divides the figure into two equal parts or two congruent parts. Such a definition is considered partially correct because, although it covers the basic idea of an axis of symmetry, the two congruent parts or two equal parts may not be symmetrical. For example, a parallelogram can be divided into two congruent triangles along any of its diagonals; however, the line of the diagonal is not necessarily an axis of symmetry of the parallelogram. The number of correct examples and nonexamples for *axis of symmetry* shows that most of these students were able to state correct examples and nonexamples. They may have been limited in their ability to precisely express these definitions in words.

The correct answer count for definitions, examples, and nonexamples by concepts are plotted in the bar chart in Figure 6.3. As you can see, in general, the definition bars are lower than the corresponding example bars and nonexample bars,

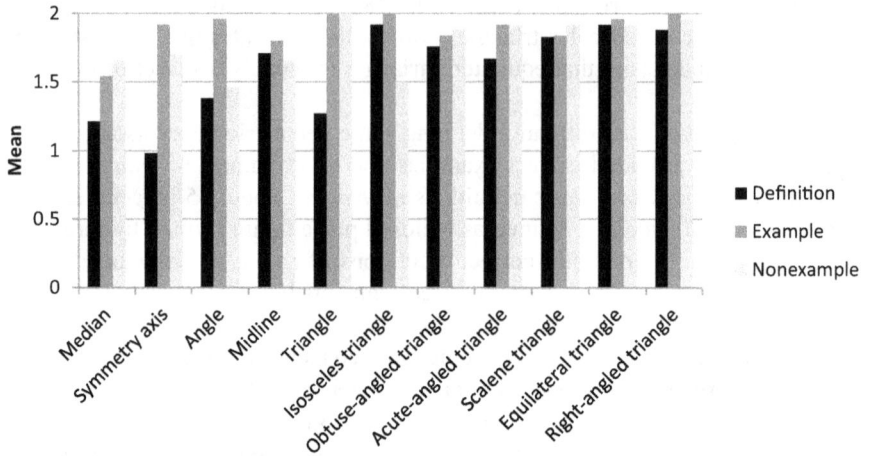

Figure 6.3 Means of definition, example, and nonexample scores of the triangle concepts.

and the example bars are higher than the corresponding nonexample bars. This suggests that the students were more skilled at providing examples and nonexamples of the concepts than in defining them, especially those that involve multiple parts, and they may be more adept at producing examples than nonexamples.

Triangles: Paper-and-Pencil Task

The P&P task (triangles) includes items testing knowledge of the individual concepts, relationships between the concepts or their properties, and applications of the concepts and relationships. The higher percentage of points reported in Table 6.8 suggests deeper comprehension of the concept as measured by the P&P task item.

The P&P task (triangles) assesses knowledge of four individual concepts, i.e. triangle, isosceles triangle, acute-angled triangle, and scalene triangle. The percentages of points earned indicate that students had considerable knowledge of *triangle* (96 percent) and *acute-angled triangle* (92 percent). The percentage of points earned for *isosceles triangle* was 84 percent, and the percentage of points earned for *scalene triangle* was only 66 percent. Item 7 in the P&P task (triangles) involves four properties of *isosceles triangle*. Most students were clear on three of the properties; however, about 44 percent of the students thought that for every isosceles triangle, one of its sides is always shorter than the other sides. They did not realise that the equilateral triangle is also an isosceles triangle and that it serves as a counter-example of the given property. This finding is consistent with the finding from the DEN task (triangles) in that some 23 percent of the 48 students used an equilateral triangle as a nonexample of an isosceles triangle. However, only five students incorrectly answered both the nonexample task and the property task, suggesting that the two tasks addressed the same misconception from different perspectives.

Item 5 in the P&P task (triangles) includes three sub-items on *scalene triangle* in the form of true-or-false statements. Most students knew that a scalene triangle could have a right angle; however, only 44 percent clearly stated that 'a scalene triangle may have two angles equal in size' was false, and 54 percent agreed that 'a scalene triangle has no axis of symmetry'. Although, over 90 percent of the students provided correct definitions, examples, and nonexamples of *scalene triangle* in the DEN task, they did not apparently fully understand *scalene triangle*, because over 33 percent of them were not clear on some of its properties.

The relationships between the concepts or their properties are examined in the P&P task (triangles) through the given statements. The students were to decide whether the statements are true, false, or uncertain. Their responses indicate that they were familiar with some of the relationships but unsure about the relationships they did not often use in solving problems. For example, Item 6(4), the statement 'if a triangle has a right angle, then two of its sides at most can be equal in length' involves the relationships between the properties of a right-angled triangle. It is a true statement, but almost 40 percent of the students indicated that it was false or were uncertain. The term 'at most' could be a factor because it requires consideration of different cases, i.e. no equal sides, two equal sides, and three equal sides, which can increase the difficulty level of an item. If the statement had been phrased more directly such as, 'if a triangle has a right angle, it cannot have three sides equal in length', more students might have correctly identified it as a true statement.

Concerning applications of the concepts, their properties, and relationships, the students' performance on the different items varied. For the items which involved direct application of definitions or properties, for example, Item 16, most students arrived at a correct answer. Ninety-eight percent of students earned the full mark for Item 16. However, when items involve multiple steps, lower percentages of students earned the full marks. For example, to answer Item 18, students had to first twice apply the property that an isosceles triangle had two equal angles. They then had to apply the property 'the angle sum of a triangle is 180°' to calculate the values of the required angles. Seventy-five percent of the students answered correctly. For Item 20, which involved more steps, fewer students (67 percent) earned the full mark.

Triangles: Correlations Among the Three Tasks

This section investigates the relationships among the CM task (triangles), the DEN task (triangles), and the P&P task (triangles). Comparative analysis reveals that the three tasks are not always consistent in terms of conclusions about the students' understanding of the concepts. Students' performed quite well in the DEN task (triangles). The percentages of acceptable responses ranged from 74 percent (median) to 97 percent (right-angled triangle and equilateral triangle). However, the percentages of acceptable links with those concepts in the CM task (triangles) were lower, from 64 percent (acute-angled triangle) to 85 percent (angle). For some concepts, e.g. median, the conclusions on the students' understanding were consistent across the two tasks. The students' performance on both tasks indicates that they had a relatively poor understanding of median. For some other concepts, e.g. acute-angled triangle, conclusions reached about students' understanding differed for the two tasks. Although the students earned a high percentage of acceptable responses (92 percent) for acute-angled triangle in the DEN task, they obtained the lowest percentage of acceptable links (64 percent) for acute-angled triangle in the CM task. About half of the links with acute-angled triangle were labelled with general linking phrases resulting in propositions such as '*acute-angled triangle* includes an *acute angle*'. The students might not have paid enough attention to using detailed linking phrases and thus may not have fully expressed their conceptual knowledge. This was the case with the other types of triangles as well.

Findings on the students' conceptual understanding, as measured by the DEN task (triangles) and the P&P task (triangles) also differed. For example, as reported in Table 6.7, the percentage of acceptable responses and the percentage of points earned for the scalene triangle were 93 percent and 66 percent, respectively. This inconsistency is reasonable because the DEN task involved definitions, examples, and nonexamples of the concept, but the relevant items in the P&P task involved the properties of the concepts. The two tasks measured different aspects of the concept.

The CM task (triangles) measures the relationships between the given concepts. The P&P task (triangles) also includes items assessing the relationships. Due to the limited time available for administering the task and the limitation of the task design, fewer relationships are examined in the P&P task (triangles) than in the CM task (triangles). Even when the same relationships are examined, the conclusions about the students' understandings of the relationships are not always consistent. For example, concerning the relationship between acute-angled triangle and scalene triangle, the percentage of points earned for the corresponding item in the P&P task (triangles) indicates that most of the students held a basic idea about the relationship between the two. Nearly 88 percent of the 48 students clearly indicated that the statement 'an acute-angled triangle cannot be a scalene triangle' is false. However, in the CM task (triangles), as reported in Table 6.8, less than 33 percent of the students specified the connection between these two concepts in their concept maps. Three students built a link from acute-angled triangle to scalene triangle, and 11 students linked scalene triangle to acute-angled triangle; all of them used the linking phrase 'may be'. Item 5(2) in the P&P task (triangle) is relevant to the relationship between scalene triangle and obtuse-angled triangle. That 93 percent of the students correctly answered that 'a scalene triangle cannot be an obtuse-angled triangle' is false suggests that most students had substantial knowledge of the relationship. In the CM task (triangle), only 19 of the 48 students (39.6 percent) linked scalene triangle and obtuse-angled triangle. Of the 19, three students labelled the link with incorrect linking phrases. They proposed that obtuse-angled triangles are all scalene triangles. However, these three students incorrectly responded to Item 5(2). None of the students who answered Item 5(2) incorrectly connected obtuse-angled triangles and scalene triangles in their concept maps. For *equilateral triangle* and *scalene triangle*, more than 50 percent of the students indicated that triangles, except the equilateral triangle, are all scalene triangles. Such misconceptions are only revealed through assessment of the seven students' concept maps: two through direct links between equilateral triangle and scalene triangle and five through the links from triangle to equilateral triangle and isosceles triangle. Two possible reasons may account for the inconsistencies in the findings between the two tasks. First, students' responses to a question may depend on how it is phrased. For example, if the statement concerning the relationship between obtuse-angled triangle and scalene triangle is given as 'an obtuse-angled triangle is a scalene triangle', the percentage of points earned for the item might change. Second, the students who were uncertain about a relationship may simply not have presented it in the concept map; hence, for those relationships, the CM task may address fewer misconceptions than traditional test items.

The relationships among the three tasks is further examined by correlating the task scores. The results of Spearman's correlation analyses are reported in Table 6.9.

The density and proposition scores are significantly correlated with the DEN-definition score. Their correlations with the example and nonexample scores are, however, low and not significant. As reported in Table 6.8, most students submitted correct examples and nonexamples of triangle concepts. This indicates that the DEN-example score and DEN nonexample score had restricted ranges and could not differentiate students' understanding of the concepts. This may account for the low correlations between the example and nonexample scores with the CM scores.

The students' CM scores are moderately and significantly correlated with their scores on the P&P task (triangles). This is explainable because the P&P task (triangles) includes items addressing the relationships between some given concepts. The same relationships are examined in the CM task (triangles). Similarly, a high correlation is observed between the students' DEN-definition scores and their

Table 6.9 Spearman's correlations between CM scores, DEN scores, and P&P scores (triangles).

		1	2	3	4	5
1	CM-density					
2	CM-proposition	.846**				
3	DEN-definition	.528**	.592**			
4	DEN-example	.252**	.270**	.574**		
5	DEN-nonexample	.329**	.335**	.456**	.439**	
6	P&P- score	.600**	.669**	.790**	.383**	.407**

** Correlation is significant at the 0.001 level (2-tailed)
* Correlation is significant at the 0.01 level (2-tailed)

Table 6.10 Direct oblimin rotated factor loadings of the scores of the CM task, the DEN task, and the P&P task (triangles).

	Factor	
	1	*2*
CM-proposition (triangles)	.964	
CM-density (triangles)	.954	
DEN-definition (triangles)	.487	.559
DEN-example (triangles)		.909
DEN-nonexample (triangles)		.740
P&P score (triangles)	.692	.304
Percentage of variance	59%	17%

Note: Only loadings numerically greater than 0.3 are included

P&P scores ($r = .790$, $p < 0.001$) because the P&P task (triangles) includes items measuring students' knowledge of individual concepts.

Factor analysis was conducted to better understand the underlying construct(s) measured by the CM task (triangles), the DEN task (triangles), and the P&P task (triangles). Since relationships are theoretically expected among the aspects of students' knowledge measured by these tasks, oblique rotation (direct oblimin) was used for the analysis. Table 6.10 reports the pattern matrix of the analysis, with factor loadings of absolute value greater than 0.3 only. This criterion is the conventional rule of thumb used to determine whether factor loadings are worth interpreting (Kline, 1994).

Task scores yielded two different factors, which accounts for 77 percent of the total variance. The first factor includes the CM scores, i.e. density and proposition score. It is related to the second component of conceptual understanding, relationships among individual concepts. The second factor includes the DEN scores and concerns the first component of conceptual understanding, individual concept. These two factors have a correlation coefficient of 0.423, which supports the use of the oblique rotation method. Note that the DEN-definition score and the P&P score load on both factors. This may be explained as follows: The students used many definition-based linking phrases in their concept maps, indicating that the DEN-definition shares common features with the CM task, especially propositions. The P&P task was designed to measure all components of conceptual understanding. Thus, it is expected that the P&P score might load on both factors, and this is partially supported by the data in Table 6.10.

Quadrilaterals

In Table 6.11, the results of students' performance of the 48 students on individual concepts in the CM task (quadrilaterals), the DEN task (quadrilaterals), and the

Table 6.11 Descriptive data of students' performance on individual concepts in the CM task (quadrilaterals), the DEN task (quadrilaterals), and the P&P task (quadrilaterals).

Concept	Definition-example-nonexample Task (DEN task)																		Paper-and-Pencil Task (P&P task)		Concept mapping Task (CM task)	
	Definition						Example					Nonexample					DEN-score	Percent of acceptable responses	Percent of points earned	Relevant item(s)	Mean number of total links	Percent of acceptable links
	No response	Wrong	Partially correct	Correct	Mean	SD	No response	Wrong	Correct	Mean	SD	No response	Wrong	Correct	Mean	SD						
Symmetry axis	2 (4.2)	10 (20.8)	19 (39.6)	17 (35.4)	1.10	0.78	2 (4.2)	2 (4.2)	44 (91.7)	1.84	0.56	5 (10.4)	1 (2.1)	42 (87.5)	1.86	0.66	225	78%	/	/	3.53	78%
Diagonal	2 (4.2)	8 (16.7)	6 (12.5)	32 (66.7)	1.46	0.82	2 (4.2)	4 (8.3)	42 (87.5)	1.76	0.66	3 (6.2)	2 (4.2)	43 (89.6)	1.80	0.62	240	83%	/	/	3.90	78%
Quadrilateral	0	4 (8.3)	26 (54.2)	18 (37.5)	1.29	0.62	0	0	48 (100)	2.00	0.00	1 (2.1)	1 (2.1)	46 (95.8)	1.92	0.40	250	87%	94%	1	4.50	86%
Rhombus	0	5 (10.4)	1 (2.1)	42 (87.5)	1.77	0.63	1 (2.1)	1	46 (95.8)	1.92	0.40	3 (6.2)	8 (16.7)	37 (77.1)	1.54	0.84	251	87%	94%	5, 6	3.79	83%
Rectangle	0	6 (12.5)	2 (4.2)	40 (83.3)	1.71	0.68	0	0	48 (100)	2.00	0.00	2 (4.2)	8 (16.7)	40 (83.3)	1.66	0.76	258	90%	88%	4, 6, 12	3.98	82%
Symmetry centre	0	1 (2.1)	1 (2.1)	46 (95.8)	1.94	0.32	2	3 (6.2)	43 (89.6)	1.80	0.62	4 (8.3)	2 (4.2)	42 (87.5)	1.76	0.66	263	91%	/	/	1.55	65%
Square	1 (2.1)	6 (12.5)	1 (2.1)	40 (83.3)	1.71	0.71	0	0	48 (100)	2.00	0.00	3 (6.3)	3 (6.3)	43 (89.5)	1.80	0.62	264	92%	98%	6, 13(1)	4.28	80%
Trapezium	1 (2.1)	5 (10.4)	2 (4.2)	40 (83.3)	1.71	0.68	0	1 (2.1)	47 (97.9)	1.96	0.28	2 (4.2)	2 (4.2)	44 (91.7)	1.84	0.56	264	92%	77%	11	2.25	75%
Parallelogram	0	2 (4.2)	1 (2.1)	45 (93.7)	1.90	0.42	1 (2.1)	1	46 (95.8)	1.92	0.40	2 (4.2)	5 (10.4)	41 (85.4)	1.70	0.72	265	92%	88%	8	4.61	86%
Isosceles trapezium	1 (2.1)	3 (6.2)	0	44 (91.6)	1.83	0.56	0	0	48 (100)	2.00	0.00	3 (6.2)	0	45 (93.7)	1.88	0.48	274	95%	/	/	2.52	69%

/: no item in the P&P task addressing the students' knowledge of the individual concept

traditional P&P task (quadrilaterals) are presented. The concepts in the table are arranged in ascending order by DEN-score.

Quadrilaterals: Definition-Example-Nonexample Task

For the DEN task (quadrilaterals), in general, the students performed well on providing examples and nonexamples of the given concepts. Their responses to the examples and nonexamples suggest that they were familiar with the prototypes of these concepts. In comparison, their performance on the definition session for the concepts was relatively poor, as can be observed by comparing the definition bars, example bars, and nonexample bars in Figure 6.4.

Most of the definition bars are lower than the corresponding example bars and nonexample bars. The students obtained the lowest DEN-score on axis of symmetry. They had the greatest difficulty in defining this concept. Two students did not respond to the definition, example, and nonexample items for this concept. Only 35 percent of the 48 students provided correct definitions of axis of symmetry, although the majority (over 85 percent) of the students provided correct examples and nonexamples. This concept was also examined for the topic triangles. The finding is consistent with findings from the DEN task (triangles). Most of the students knew what axis of symmetry is; they simply had difficulty defining it accurately.

By contrast, they performed well on the definition session for centre of symmetry, a concept similar to axis of symmetry. This is perhaps because the definition of centre of symmetry is less cognitively demanding than the definition of axis

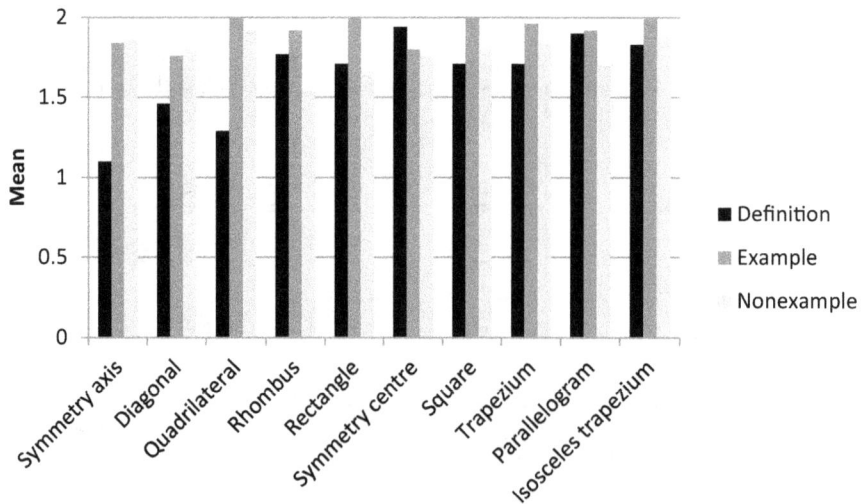

Figure 6.4 Means of definition, example, and nonexample scores of the quadrilateral concepts.

of symmetry. The centre of symmetry of a plane figure is defined as a fixed point of the figure with respect to which the figure shall coincide with itself after rotating 180°, while the axis of symmetry is defined as a line in a plane figure which divides the figure into two parts such that one part, when folded over along the axis, shall coincide with the other part. As long as the student included 'rotating 180°' and 'coincide or overlap with itself', the definition of *centre of symmetry* was considered correct.

The students did not perform well on the definition session for quadrilateral. More than 50 percent provided partially correct definitions, although almost all 48 students provided correct examples and nonexamples. This is quite similar to the finding for the definition session for *triangle*, perhaps because these two concepts are defined in a similar manner. *Triangle* is defined as a closed plane figure formed by three line segments; *quadrilateral* is defined as a closed plane figure formed by four line segments that do not intersect. Students' partially correct definitions include that *quadrilateral* is a figure with four sides but do not emphasise that it should be a closed figure or that the four sides should be joined end to end.

The students obtained high DEN-scores on special types of quadrilaterals, especially *parallelogram* and *isosceles trapezium*. During interviews (see Chapter 7), students confirmed that they were quite familiar with these concepts because they were newly learned and the terms used for these concepts (in Chinese) specify their properties, which helps students in memorizing the concepts and differentiating them from each other. For example, parallel (平行), the prefix of *parallelogram* (平行四边形) facilitated their remembering that a quadrilateral has parallel sides.

Quadrilaterals: Paper-and-Pencil Task

The P&P task (quadrilaterals) measures students' knowledge of individual concepts, their relationships, and applications of the concepts and relationships.

It includes items addressing students' knowledge of the following six concepts, quadrilateral, rhombus, rectangle, trapezium, square, and parallelogram. The percentage of points students earned for each concept indicates that they generally had substantial knowledge of these concepts. Comparatively, the percentage of points earned for trapezium (77 percent) is lower than the percentages earned for the other five concepts. Item 11, the relevant item, is a multiple-choice question with five options. Each option describes a quadrilateral with a certain property. Students are to decide whether or not a given quadrilateral could be a trapezium. Errors were found. Some 21 percent of the students indicated that 'a quadrilateral with one pair of opposite sides equal in length' cannot be a trapezium, another 21 percent indicated that 'a quadrilateral with two equal diagonals' cannot be a trapezium, and almost 13 percent indicated that 'a quadrilateral with two pairs of equal sides' cannot be a trapezium. A counter-example to these errors is the isosceles trapezium. Isosceles trapezium is a trapezium with a pair of non-parallel, equal sides and equal diagonals. In addition, almost 20 percent of the students indicated that 'a quadrilateral with two right angles' cannot be a trapezium.

A counter-example to this error is the right-angled trapezium, a trapezium with two right angles. Some 17 percent of the students thought that 'a quadrilateral with two perpendicular diagonals' cannot be a trapezium. This is also incorrect. The errors found here were not detected in the DEN task (quadrilaterals) because the task involves only definitions, examples, and nonexamples of the *trapeziums*, while items from the P&P task examine the properties of trapeziums.

The findings from the P&P task (quadrilaterals) indicate that the students had a good understanding of the relationships between the given concepts and their properties. For example, Item 2 examines understanding of the relationships between the properties of different rectangles, squares, and parallelograms. Most students were familiar with the set-subset relationships between the concepts. The percentage of points earned for this item is 85 percent. Students seemed to have had some difficulty in working with the relationships between the properties of diagonals, sides, angles, and axes of symmetry. For example, for Item 7, students were to decide which of the five given statements is(are) always true for any four-sided closed figure. Statements such as 'if its diagonals bisect each other, then all its sides are equal in length' do not identify specific quadrilaterals. Some 45.8 percent did not answer correctly. To answer such a question, students may need to first think of specific quadrilaterals that satisfy the condition(s), then judge whether the quadrilateral had the property in the deduction.

Concerning the applications of the concepts and their relationships, the students' responses to the relevant items in the P&P task (quadrilaterals) indicate that they were able to effectively apply their knowledge in solving routine and simple problems, such as finding the values of unknowns in a quadrilateral with given conditions (Item 18) and simple proof (Item 19). However, they were relatively weak in reasoning and in solving problems that require the application of multiple properties. For example, to answer Item 21, one must make logical inferences based on the given conditions. Substantial understanding of the properties of parallelogram, rhombus, and rectangles and flexible application of those properties are important for achieving the final answer in such cases. Only 60 percent of the students earned the full mark for this item.

Quadrilaterals: Correlations Among the Three Tasks

This section reports on the relationships among the CM task (quadrilaterals), the DEN task (quadrilaterals), and the P&P task (quadrilaterals). Descriptive analyses in previous sections indicate that the inferences drawn about the students' understanding of the given concepts were not always consistent across the three tasks. Each tasks appears to have its own advantages in measuring conceptual understanding.

The percentages of acceptable responses of the concepts in the DEN task (quadrilaterals) range from 78 percent (axis of symmetry) to 95 percent (isosceles trapezium). The percentages of acceptable links for the concepts in the CM task (quadrilaterals) range from 65 percent (centre of symmetry) to 86 percent (quadrilateral and parallelogram). The percentages of acceptable links with concepts is generally lower than the percentages of acceptable responses, suggesting that the students' knowledge of definitions, examples, and nonexamples of the concepts may have been more substantial than their knowledge of relationships among the concepts. For example, over 90 percent of the students correctly defined isosceles trapezium and provided correct examples and nonexamples, but the percentage of acceptable links with isosceles trapezium is only 69 percent. A close look at the students' concept maps reveals that the linking phrases are mostly general or partially correct. This accounts for the relatively low percentages of acceptable links with the concepts. The students indicated an incorrect understanding of some properties of isosceles trapezium in the CM task. For example, they constructed propositions such as '*isosceles trapezium* had a *centre of symmetry*' and '*isosceles trapezium* has perpendicular *diagonals*'. Such misconceptions are not revealed by the DEN task.

The CM task (quadrilaterals) assesses knowledge of the relationships between the given concepts. Compared with the CM task, other than relationships between concepts, the P&P task (quadrilaterals) measures the other two target aspects of the students' conceptual understanding, i.e. knowledge of individual concepts and their operations. Even where they measure same aspect, i.e. relationships between individual concepts, they each measure it from a different perspective. Consider the case of the relationship between *parallelogram* and *rectangle*. Item 3 in the P&P task is a multiple-choice question. Students choose from four properties satisfying the condition that all rectangles have but that some parallelograms do not have. In the CM task, students mostly mapped the connections between *parallelogram* and *rectangle* with linking phrases like 'with a right angle' and 'with equal diagonals'. They addressed these relationships differently.

The P&P task (quadrilaterals) includes items measuring understanding of the individual concepts trapezium, square, rectangle, rhombus, quadrilateral, and parallelogram (see Table 6.11). The percentages of points students earned for these concepts in the P&P task and the percentages of acceptable links in the DEN task are generally consistent, except for the concept *trapezium*.

The relationships among the three tasks is further examined by correlating their task scores. The results of the Spearman's correlational analyses are reported in Table 6.12. Moderate and statistically significant correlations are observed between the CM scores (density and proposition scores) and DEN-definition score. Their correlations with the DEN-example score and the DEN-nonexample score are low (r ranges from -.136 to .223). Similarly, low correlations are observed between the P&P and DEN-example scores and the DEN-nonexample score. There are two possible reasons for the low correlations. First, the DEN-example and DEN-nonexample scores have restricted ranges due to the students' facility with providing examples and nonexamples of the geometric concepts. Second, the DEN-example and DEN-nonexample only involve individual concepts. The CM task is mainly about relationships between the concepts; few students included examples or nonexamples of the concepts in their maps. In the P&P task (quadrilaterals), only one item directly elicits examples of *parallelogram*.

Table 6.12 Spearman's correlations between CM scores, DEN scores, and P&P scores (quadrilaterals).

		1	2	3	4	5
1	CM-density					
2	CM-proposition	.871**				
3	DEN-definition	.429**	.572**			
4	DEN-example	.107**	.223**	.606**		
5	DEN-nonexample	-.136***	.117**	.317**	.674**	
6	P&P- score	.585**	.694**	.719**	.344**	.100**

** Correlation is significant at the 0.001 level (2-tailed)
* Correlation is significant at the 0.01 level (2-tailed)

Table 6.13 Direct oblimin rotated factor loadings of the scores of the CM task, the DEN task, and the P&P task (quadrilaterals).

	Factor	
	1	2
CM-proposition (quadrilaterals)	.934	
CM-density (quadrilaterals)	.935	
DEN-definition (quadrilaterals)	.636	.492
DEN-example (quadrilaterals)		.888
DEN-nonexample (quadrilaterals)		.907
P&P score (quadrilaterals)	.826	
% of variance	53%	28%

Note: Only loadings numerically greater than 0.3 are included

The CM-proposition score has a higher correlation (r = .694) with the P&P score than CM-density (r = .585). This is consistent with the findings from the correlational analyses of the other three topics, suggesting that the proposition score is a better indicator of conceptual knowledge than the density of the concept maps. In addition to its higher correlations with the DEN scores and the P&P score for the four topics, propositions provide more information on students' understanding of the concepts than density. This might be why more studies in the literature consider the proposition score of concept maps rather than the densities.

Results of factor analysis conducted with the task scores are reported in Table 6.13; the task scores yield two different factors accounting for 81 percent of the variance.

These two factors have a correlation coefficient of 0.190. The first factor covers the density and proposition scores for the CM task and the P&P task score. Since the P&P task includes the items addressing the relationships between different concepts and between different properties of the same concepts, this factor may be more related to the relations among individual concepts component of conceptual understanding. The second factor covers the DEN-example and DEN-nonexample scores. It involves the individual concepts component of students' conceptual understanding of the quadrilateral concepts. The DEN-definition score loads on both factors. This finding is similar to that of the factor analysis for the topic triangles, except that the P&P score (quadrilaterals) loads on only one factor, while the P&P score (triangles) loads on both factors. The P&P task (quadrilaterals) has fewer items on individual concepts but more items on relationships between concepts than the P&P task (triangles), which may be affecting the results of the factor analysis.

Differences Among Mathematical Topics

These analyses suggest that the relationships among the CM tasks, the DEN tasks, and the P&P tasks differ by topic. These differences can be easily observed from the results of the correlational analyses and factor analyses. The factor analyses suggest that for the two algebraic topics, the three tasks measured the same underlying construct, while for the two geometric topics, the three tasks might actually measure different things.

Two possible explanations for these differences are as follows: First, conceptual understanding in algebra and in geometry may have, theoretically, different structures. At the secondary school level, algebra mostly involves symbolic mathematical statements and basic algebraic operations, while geometry primarily involves size, position, relations, and their transformations of geometric shapes (Ministry of Education, China, 2011). These are two genuinely different domains of mathematics. Second, the DEN-example scores and the DEN-nonexample scores for the two geometric topics present little variation. As reported in Tables 6.2, 6.5, 6.8, and 6.11, the given geometric concepts generally yielded higher means and lower standard deviations than the algebraic concepts in this study. In addition, the design of the P&P tasks for the algebraic topics and the geometric topics do not follow uniform criteria; for example, the percentages of items for each of the P&P tasks should be on individual concepts, relationships among individual concepts, and relationships among concepts and operations, respectively. This might be a limitation of the DEN tasks and the P&P tasks which prevented clearly settling the task scores in factor analyses. Accordingly, improvements to the design of the DEN tasks and P&P tasks are needed. Further research may be needed to check for possible topical differences regarding the use of concept mapping as an assessment technique of students' conceptual understanding in mathematics.

Summary

This chapter reports on students' performance on the DEN tasks and the traditional P&P tasks for four different topics, triangle, quadrilateral, algebraic expression, and equation, by comparing it with their performance in the CM tasks. Both qualitative analyses of descriptions and quantitative correlational analyses are applied.

The three types of tasks emphasise different aspects of understanding of the concepts. Compared to the DEN tasks and the P&P tasks, the CM tasks more directly address the relationships between the concepts. They provide information on students' understanding of the hierarchy of the given concepts. Such information cannot be elicited from students through the DEN tasks and the P&P tasks. Certainly, the CM tasks also have their limitations. In contrast to the DEN tasks, the CM tasks were limited in eliciting information on the students' knowledge of the individual concepts. Few students included examples or nonexamples of the given concepts in their concept maps. Unlike the P&P tasks, the CM tasks are limited in testing knowledge of relationships between concepts and operations. No information on students' level of facility applying their knowledge of individual concepts and relationships was elicitable from the concept maps.

Correlations between the CM-density and the CM-proposition scores with the DEN-definition scores and the P&P scores are positive and statistically significant across the four topics. Differences are detected between the geometric topics and the algebraic topics. For the two geometric topics, triangle and quadrilateral, the densities and the proposition scores have low correlation coefficients with the DEN-example scores and the DEN-nonexample scores; for the two algebraic topics, the correlations are higher and are all statistically significant. Factor analyses also reveals that the task scores for the two algebraic topics led to a single factor, which supports the concurrent validity of the concept map as an assessment of conceptual understanding. However, the task scores for the two geometric topics yielded two different factors, suggesting that the tasks emphasise only particular components of conceptual understanding.

In addition to topic differences, the correlational analyses indicate that the CM-proposition scores are consistently more strongly correlated with the DEN-definition scores and the P&P scores than with the CM-densities. This suggests that the aspects of the students' conceptual understanding tapped by the CM-proposition share more similarities with the aspects measured by the DEN-definition and the P&P task than with CM-density. This is consistent with findings from the literature that proposition scores are used more often than density as an indicator of students' performance in concept maps.

In summary, each of the three task types emphasise and address particular components of conceptual understanding of given concepts. Together, they can provide deeper insight into students' conceptual understanding. Further research may be needed to examine the possible topical differences and validity issues with concept mapping as a technique for assessment of conceptual understanding in mathematics.

Note

The comparison between concept mapping and the two school tests on the topic *triangle* has been published as a journal papers: Jin, H., and Wong, K. Y. (2021) 'Complementary measures of conceptual understanding: a case about triangle concepts', *Mathematics Education Research Journal*, online-first version, doi.org/10.1007/s13394–021–00381-y

7 Attitudes Toward Concept Mapping

Study Design

The concept map has been advocated as an effective tool for teaching, learning, and assessment of conceptual understanding in science education. Although the term *concept map* might not be new to many students and mathematics teachers around the world, few are familiar with its use, and fewer apply it in educational environments. How students and mathematics teachers feel about concept mapping and how likely they are and how willing they are to use and incorporate it in learning—teaching of mathematics are issues of interest.

Participants

Students

The students were the class ($n = 48$, aged 13–14 years) who underwent comprehensive training on concept mapping and participated in the CMS-training and the four CM tasks, as described in Chapters 4, 5, and 6. With these experiences, they gained substantial familiarity with concept mapping.

Teachers

Both prospective and in-service teachers participated. The prospective teachers comprised 173 undergraduates and 176 master students who majored in mathematics education at three universities in Jiangsu and Zhejiang, China. The in-service teachers comprised 55 mathematics teachers from different schools in Jiangsu, China. They were attending summer training courses at the researcher's university. Before data collection, a ten-minute introductory session was conducted with prospective and in-service teachers which covered what concept mapping is, how to construct a concept map, and its application as a teaching/learning strategy and an assessment device. Table 7.1 presents the background information of the teachers, including their gender, age group, grade (for prospective teachers), and school (for in-service teachers). As shown in the table, the number of female teachers is more than twice the number of male teachers. This is consistent with the general situation in China in that there are markedly more female teachers than male

DOI: 10.4324/9781003269373-7

Table 7.1 Background information of the teachers.

			No. of participants	Percent
Prospective Teachers (university students) (n=173)	Gender	Male	68	39.3%
		Female	105	60.7%
	Grade	Year 2	36	20.8%
		Year 3	137	79.2%
	Age group	Under 25 years old	173	100%
Prospective Teachers (master students) (n=176)	Gender	Male	40	22.7%
		Female	136	77.3%
	Grade	Year 1	72	40.9%
		Year 2	65	36.9%
		Year 3	39	22.2%
	Age group	Under 25 years old	135	76.7%
		26 to 30 years old	41	23.3%
In-service Teachers	Gender	Male	13	23.6%
		Female	42	76.4%
	School	Primary	24	43.6%
		Secondary	13	23.6%
		High school	12	21.8%
		College or University	6	10.9%
	Age group	Under 25 years old	6	10.9%
		26 to 30 years old	40	72.7%
		31 to 35 years old	9	16.4%

Source: Adapted from Jin, Lu and Zhong, 2015, p. 596

teachers, especially in kindergarten and primary schools (e.g. Wang and Tang, 2004; Li, 2005).

Instruments: Questionnaires, Interview, and Open-Ended Tasks

Instruments for Students

Since there is no validated questionnaire in the literature addressing students' and teachers' attitudes toward concept mapping in mathematics, two questionnaires, one for students and one for teachers, were self-designed by the present research with reference to the existing literature. The students' questionnaire is called the Attitudes Toward Concept Mapping Questionnaire (ATCMQ). It was designed with reference to Mohamed's (1993) attitudes toward concept mapping in science questionnaire. Some items were adapted from Kankkunen's (2001) study, in which students' opinions on concept mapping were gathered through inquiry and interviews.

The ATCMQ comprises 35 items, 11 of which are negatively stated. The use of negatively stated items strongly encourages students to read and evaluate the items carefully. The 35 items are divided among the following six aspects:

1   *Making inferences from a given concept map.* Five items on this aspect are included. It reflects students' perceptions of making inferences from given

concept maps. A high score indicates a high perceived ability to make infer-ences from given concept maps.

2   *Ease of constructing a concept map.* Five items on this aspect are included. It reflects students' perceived ease in constructing a concept map. A high score indicates high perceived ease in concept mapping.

3   *Confidence in concept mapping proficiency.* Six items on this aspect are included. It reflects students' confidence in their concept mapping profi-ciency. A high score indicates high confidence in their ability to construct concept maps.

4   *Enjoyment of concept mapping.* Six items on this aspect are included. It reflects the pleasure students derive from engaging in concept mapping activ-ities. A high score indicates high enjoyment of concept mapping.

5   *Usefulness of concept mapping.* Eight items on this aspect are included. It reflects students' views regarding the value of using concept mapping in mathematics. A high score indicates high perceived usefulness of concept mapping.

6   *Preference for using concept map for further study.* Five items on this aspect are included. It reflects students' preference for using concept mapping in their further study of mathematics. A high score indicates high preference for using concept mapping in further study.

Table 7.2 overviews the aspects and the corresponding items. Unlike Moham-ed's (1993) and Kankkunen's (2001) questionnaires, the ATCMQ includes items addressing students' perceptions of their concept mapping skills. The five items addressing making inferences from a given concept map were designed by the researcher. The students received sample concept maps in the training section. Ability to read information from a concept map is a first step toward concept mapping. Ease of constructing a concept map focuses on specific mapping skills,

Table 7.2 Aspects of the Attitudes Toward Concept Mapping Questionnaire and its cor-responding item numbers.

No.	Aspects	Item Number	Total No.
1	Making inference from given concept maps (Inferences)	1, 2, 5, 7, 24*	5
2	Ease of constructing concept map (Ease)	9, 15*, 19, 31*, 34*	5
3	Confidence with concept mapping (Confidence)	6, 11*, 22, 28*, 33*, 35	6
4	Enjoyment of concept mapping (Enjoyment)	4*, 13*, 16, 20, 23, 26*	6
5	Usefulness of concept map (Usefulness)	3, 8, 10, 17, 21*, 27, 29, 32	8
6	Preference of using concept map for further study (Preference)	12, 14, 18, 25, 30	5
Total number of items			35

* The asterisk indicates that the item was negatively worded

e.g. 'I find it difficult to add accurate linking phrases' (Item 15) and 'I find it easy to determine the hierarchy among a list of given concepts' (Item 19). Item 13 is taken directly from Mohamed's questionnaire, Items 12, 14, 18, and 22 are modified from Mohamed's questionnaire, and Items 8 and 20 are modified from Kankkunen's questionnaires. The items were translated into Chinese. The face validity of the Chinese version of the questionnaire was verified by two mathematics teachers in an experimental school. They were asked to ensure that the students would likely understand the items.

In the pilot study for the ATCMQ, students rated the items on a five-point Likert scale, and some students selected the neutral option for many of the items. This may be due to their not having carefully considered their own attitudes. To encourage them to fully express their attitudes, for the finalised ATCMQ administered to participants in the main study, I adopted a six-point Likert scale, from 1 = Strongly Disagree (SD) to 6 = Strongly Agree (SA). In this way students were required to take a stand. Analyses of the data from the pilot study and comments from the students in both the preliminary and pilot study necessitated modifications to several items. The finalised ATCMQ is provided in Appendix E.

Interviews are scheduled after the ATCMQ to further explore students' thinking process and strategies during concept mapping and probe in-depth their attitudes toward concept mapping. The interview questions are listed here:

1   Given concepts A, B, C, D, and E, please describe how you construct a concept map using these given concepts.
2   Which one do you find easier: mapping with algebraic concepts or with geometric concepts? Why?
3   What kind of concept maps do you find is a good one?
4   Among the three types of tasks, statement transformation, simple free association, and extended free association, which one do you think is the easiest? Why?
5   In which aspect(s) do you think concept mapping is helpful? (a) Review, (b) Memorization, (c) Problem solving, (d) Understanding concepts, (e) Others: (please specify) _____, or (f) No use at all. Please indicate your agreement by ticking one or more of the choices and provide your explanations below.
6   Can you summarize your major achievements during this concept mapping period?

For students who were not interviewed, these questions were assigned as an open-ended written task.

Instruments for Teachers

The teachers' questionnaire was modified from the students' ATCMQ. It focuses on three aspects: interest in concept mapping, appreciation of usefulness of concept mapping, and willingness to use concept mapping. The face validity was verified by three mathematics educators at normal universities. It includes 22 items, six of which are negatively worded. Participants rated items on a five-point Likert scale, from 1 = Strongly Disagree (SD) to 5 = Strongly Agree (SA). Reverse scoring was applied to the negatively stated items.

The interviews were guided by the following six questions: the first three are about the feasibility of using concept maps in school settings; the fourth and fifth questions are relevant to teachers' willingness to use concept maps in the class-room, and the sixth question concerns students' attitudes about concept maps from the teachers' perspectives.

1   How do you prefer to introduce concept mapping, through concentrated training or by providing reference materials for self-study?
2   How feasible is the use the concept mapping in mathematics education?
3   Do you think concept mapping is more suitable for teaching, learning, or assessment?
4   Would you use concept mapping in teaching?
5   Would you like to introduce concept mapping to your students and encourage them to use it for learning mathematics?
6   From your perspective, what will the students think about the concept map?

For some other teachers who were not interviewed, these questions were assigned as an open-ended written task.

Data Collection

Data on the students and teachers' attitudes on concept mapping were collected through the questionnaires, interviews, and open-ended written tasks.

After the ATCMQ, 12 students were selected for the interview with stratification sampling based on their school mathematics achievements. The interviews were conducted one-to-one. Each interview took about 15 to 30 minutes and it was audio-taped. The six interview questions were assigned to the remaining 36 students as an open-ended written task at the end of the main study. They were given 30 minutes to write down their answers. The interviews and the open-ended written task together could inform and interpret the results of the CM tasks and the students' responses in the ATCMQ.

After the teachers' questionnaire, four prospective teachers and four in-service teachers were randomly selected and interviewed so as to obtain detailed information on their attitudes toward using the concept map in mathematics in general. The same questions were assigned to another 11 prospective teachers as an open-ended task after they developed a lesson plan incorporating concept map in different stages of teaching. Their responses to the questions were sent back to the researcher by email.

Table 7.3 An overall view of the Attitudes Toward Concept Mapping Questionnaire and the students' responses.

Aspects	No. of items	Cronbach's α	Mean	SD
Usefulness	8	0.724	5.06	0.20
Inferences	5	0.775	4.71	0.30
Preference	5	0.867	4.37	0.20
Enjoyment	6	0.830	4.22	0.35
Confidence	6	0.724	4.16	0.88
Ease	5	0.339	3.43	0.87

Table 7.4 Aspects of the Attitudes Toward Concept Mapping Questionnaire and its corresponding item numbers.

No.	Aspects	Cronbach's α
1	Interest on concept map	0.751
2	Appreciation of the usefulness of concept map	0.713
3	Willingness to use concept map	0.765

Source: Adapted from Jin, Lu, and Zhong, 2015, p. 599

Data Analysis

Students' responses to the ATCMQ are discussed later using descriptive statistics. The internal consistency of the aspects is examined using Cronbach's alpha. Table 7.3 presents an overview of the aspects, the number of items, Cronbach's α, and the mean scores of the students' responses for each aspect. The mean scores are calculated by averaging the means of the items, with the scores of the negatively worded items being reversed. For the six-point scale, a mean score of 4.5 is the cut-off point for indication of moderate to high levels of agreement.

The Cronbach's alphas of the aspects, except for ease of constructing concept map, indicate their acceptable internal consistency (Cronbach's α > .70; see Hair et al., 1998, p. 730). Ease of constructing concept map had low internal consistency (Cronbach's α = .339), suggesting that the items do not belong, conceptually, to the same aspect. The mean scores of the other aspects ranged from 4.16 to 5.06, indicating various levels of agreement.

Table 7.4 presents an overview of the aspects of the teachers' questionnaire and their corresponding Cronbach's α (based on standardised items). The Cronbach's alphas of the three aspects indicate their acceptable internal consistency.

Students' Attitudes Toward Concept Mapping

Usefulness

This aspect involves students' perceptions of the usefulness of concept mapping. Table 7.5 reports the distribution of their responses to the eight items related to

Table 7.5 Frequencies, percentages (in parentheses), means, and SD of responses to *use-fulness* of concept map.

Items	Frequency						Mean	SD
	SD	D	LD	LA	A	SA		
17. Using concept map, one can clearly describe relationships between mathematical concepts.	0	0	0	8 (16.7)	16 (33.3)	24 (50.0)	5.33	.75
8. Concept map can reflect what I understand about the concepts.	0	0	3 (6.4)	5 (10.6)	17 (36.2)	22 (46.8)	5.23	.89
21*. Concept mapping is a waste of time.	21 (44.7)	11 (23.4)	14 (29.8)	1 (2.1)	0	0	5.11	.91
3. Concept mapping is helpful for understanding mathematics concepts.	0	1 (2.1)	1 (2.1)	10 (20.8)	16 (33.3)	20 (41.7)	5.10	.95
29. After doing a concept map, I can see more clearly how the concepts are related.	0	0	2 (4.2)	8 (16.7)	22 (45.8)	16 (33.3)	5.08	.82
10. It is fair to judge our conceptual achievement according to our performance in a concept map.	0	1 (2.1)	3 (6.4)	9 (19.1)	16 (34.0)	18 (38.3)	5.00	1.02
32. I believe that concept mapping is useful for our study.	1 (2.1)	1 (2.1)	1 (2.1)	12 (25.0)	13 (27.1)	20 (41.7)	4.98	1.14
27. I can come up with new ideas when doing concept mapping.	0	3 (6.3)	2 (4.2)	11 (22.9)	24 (50.0)	8 (16.7)	4.67	1.02

this aspect in the questionnaire. SD, D, LD, LA, A, and SA in the first row of the table represent Strongly Disagree, Disagree, Slightly Disagree, Slightly Agree, Agree, and Strongly Agree, respectively. The top figures in the table refer to the frequencies, and the bottom figures in parentheses are percentages. The items are ranked in descending order according to their means. The students expressed significantly positive attitudes regarding the usefulness of concept mapping. All the means of these items were greater than 4.50, with the means of negatively worded items reversed.

All 48 students agreed with the statement that one can clearly describe the relationships among mathematical concepts using the concept map (Item 17). With only one exception, they disagreed with the statement that concept mapping is a waste of time (Item 21). Items 8 and 10 gauged students' attitudes toward the use of the concept mapping as an assessment technique; over 90 percent of the students expressed positive attitudes for this use. They generally perceived that concept mapping is suitable for reflecting their understanding of mathematical concepts (Item 8) and considered it fair to evaluate their conceptual achievement according to their performance on a concept mapping task (Item 10). Although students were able to add more propositions when prompted during the CM tasks, from the students' perspective, concept mapping is valuable as an assessment technique. For example, over 75 percent of the students indicated that concept mapping helped them both in remembering the concepts learned and in better understanding new concepts.

The students also affirmed that they could benefit from concept mapping (Items 3, 27, and 29). More than 90 percent of the students agreed with the statement that concept mapping is helpful for understanding mathematics concepts (Item 3) and could more clearly see how concepts are related after constructing a concept map (Item 29). For Item 27, whether they generate new ideas while concept mapping, the eight students who chose 'strongly agree' comprised both high and low achieving students, suggesting that students at various academic levels could benefit from concept mapping. Together these findings present a vote of confidence from students for developing the concept map as a tool assessing conceptual understanding in mathematics and for further exploring it as a learning tool.

Making Inferences From Given Concept Maps

This aspect involves students' perceptions of their proficiency in making inferences from given concept maps. The distribution of the students' responses to the items is reported in Table 7.6. The mean for four out of the five items is above the 4.5 cut-off point after reverse scoring of the negative item, and the mean for one item is 4.35, suggesting a moderate level of agreement with the statements by most students. The students generally responded positively regarding their proficiency in making inferences from concept maps.

Item 1, the item with which the most students strongly agreed, concerns their ability to write statements about a given concept map. The majority of students

Table 7.6 Frequencies, percentages (in parentheses), means, and SD of responses to *making inferences from* given concept maps.

Items	Frequency						Mean	SD
	SD	D	LD	LA	A	SA		
1. I can write statements from a given concept map.	0	0	0	9 (18.8)	28 (58.3)	11 (22.9)	5.04	0.65
2. I can understand the hierarchy among concepts in a given map.	0	0	4 (8.3)	7 (14.6)	22 (45.8)	15 (31.3)	5.00	0.90
24*. I have no idea about how to read a concept map.	10 (20.8)	22 (45.8)	9 (18.8)	3 (6.3)	4 (8.3)	0	4.65	1.14
5. I can judge whether the linking phrases in a given map are accurate.	1 (2.1)	2 (4.2)	6 (12.5)	8 (16.7)	25 (52.1)	6 (12.5)	4.50	1.13
7. I can distinguish a good map from a bad one.	1 (2.1)	4 (8.3)	4 (8.3)	13 (27.1)	20 (41.7)	6 (12.5)	4.35	1.19

might have agreed or strongly agreed with this item because writing statements is quite straightforward, as long as one knows how to read the propositions. This is the most basic skill required for using concept maps for learning, and all the students felt they possessed the skill.

Understanding the hierarchy among mathematical concepts is difficult because it requires recognising the whole-part and set-subset relationships among the concepts. I did not stress or extensively explain hierarchy, nor did I set strict requirements on how to present concepts hierarchically. Thus, students may have had their own ideas about the meaning of hierarchy. Nevertheless, most students responded positively, with only three students somewhat disagreeing with the statement that they could understand the hierarchy of concepts in a concept map (Item 2).

Most students expressed moderate agreement that they were able to read information from concept maps (Item 24). They were also quite positive about their proficiency in judging the accuracy of linking phrases in a map (Item 5). However, large standard deviations were calculated for those two items, suggesting that the students had different perceptions about these competencies.

The students were less confident in their abilities to distinguish a high-quality concept map from a low quality one (Item 7); some 50 percent of the students somewhat agreed with the statement in Item 7. They may not have been sure of what a high-quality concept map is. The requirements set for quality concept maps were introduced only during the training stage. Although I introduced numerous examples of well-constructed and poorly constructed concept maps to the readers, I did not present any such examples to students during the practice stage. Accordingly, a lack of concrete examples of high- and low-quality concept maps and practice in distinguishing between the two may be a factor.

Preference for Using Concept Maps for Further Study of Mathematics

Through this aspect students' preferences for using concept maps in their further study of mathematics is measured. The distribution of students' responses is reported in Table 7.7.

In general, the students' responded positively. The means for the first two items (Items 12 and 18) are above the cut-off point of 4.5. The majority of students indicated their preference for using concept mapping in their further studies in general and for mathematics in particular, but few expressed strong agreement. Items 14 and 25 asked whether students preferred that their teachers use concept mapping in teaching. More than 40 percent of the students selected somewhat agree. The students seemed to want to use concept map and appreciated its usefulness but were uncertain whether they would continue to use the tool on their own. Moreover, in informal talks, some students stated that they generally considered problem solving to be the most important factor in learning mathematics and that they did not think concept mapping is helpful for problem solving. This may be a factor in their lack of strong preference for using the concept map as a tool in their further study of mathematics.

Table 7.7 Frequencies, percentages (in parentheses), means, and SD of responses to *preference of using* concept map for further study.

Items	Frequency						Mean	SD
	SD	D	LD	LA	A	SA		
12. I will try to use concept maps in my further study.	0	2 (4.2)	2 (4.2)	15 (31.3)	22 (45.8)	7 (14.6)	4.63	.94
18. I'd like to use concept maps in mathematics.	0	5 (10.4)	1 (2.1)	14 (29.2)	21 (43.8)	7 (14.6)	4.50	1.11
30. I would like to have more concept mapping activities in my mathematics lessons.	2 (4.2)	1 (2.1)	7 (14.6)	13 (27.1)	18 (37.5)	7 (14.6)	4.35	1.21
13. I hope our teacher can use concept maps to teach us mathematical concepts.	1 (2.1)	2 (4.3)	5 (10.6)	19 (40.4)	15 (31.9)	5 (10.6)	4.27	1.08
25. I can learn mathematics better if my teacher use concept maps for teaching.	0	4 (8.3)	8 (16.7)	20 (41.7)	11 (22.9)	5 (10.4)	4.10	1.08

Moreover, most of the students who indicated lack of preference are female. Five out of the six students who expressed disagreement with Item 18 are also female, perhaps suggesting that male students have a higher preference for using concept map in their future studies. Gender differences need to be studied further.

Enjoyment of Concept Mapping

This aspect refers to whether students enjoy concept mapping. Table 7.8 shows the distribution of their responses to the six items on *enjoyment*. In general, most students indicated moderate enjoyment of concept mapping.

The students expressed the strongest disagreement for the item 'I find concept mapping boring' (Item 13). Only seven of the 48 students selected somewhat agree for this item; the majority (85 percent) selected either somewhat disagree or strongly disagree. More than 80 percent of the students agreed with the statement that concept mapping is interesting (Item 20), and some 75 percent agreed with the statement that they liked spending time on concept mapping (Item 16). The students' consistently positive responses to the items suggest that they did enjoy concept mapping.

Items 23 and 26 address students' possible anxiety toward concept mapping. Though most of the students admitted that concept mapping is challenging (Item 34, ease of constructing concept map), 70 percent agreed with the statement that

Table 7.8 Frequencies, percentages (in parentheses), means, and SD of responses to *enjoyment* of concept mapping.

Items	Frequency						Mean	SD
	SD	D	LD	LA	A	SA		
13*. I find concept mapping boring.	14 (29.2)	18 (37.5)	9 (18.8)	7 (14.6)	0	0	4.81	1.02
20. Concept mapping is interesting to me.	0	3	5 (6.4)	14 (10.6)	19 (29.8)	6 (40.4) (12.8)	4.43	1.06
26*. I feel anxious when I am asked to construct a concept map.	12 (25.0)	13 (27.1)	7 (14.6)	6 (12.5)	9 (18.8)	1 (2.1)	4.21	1.53
16. I like spending time on concept mapping.	1 (2.1)	0	12 (25.0)	18 (37.5)	14 (29.2)	3 (6.3)	4.10	.99
4*. Concept mapping is time-consuming.	8 (16.7)	10 (20.8)	9 (18.8)	14 (29.2)	6 (12.5)	1 (2.1)	3.94	1.38
23. I feel at ease when I heard the word 'concept map'.	0	7	7 (14.6)	21 (14.6)	12 (43.8)	1 (25.0) (2.1)	3.85	1.03

they felt at ease when they heard the phrase 'concept map' (Item 23), and some 65 percent disagreed with the statement that they felt anxious when they were asked to construct a concept map (Item 26). Knowing that their concept map scores would not be used to judge their school performance may have been a factor in this.

Students' views on whether concept mapping is time-consuming varied. Of the eight students who selected strongly disagree for this item, six were male students; of the seven who selected agree or strongly agree, only two were male students. It seems that the female students in this study would have liked more time to construct concept maps, compared to the male students. This study did not aim to explore gender differences, but these results suggest that exploration of gender differences in concept mapping in mathematics could be fruitful. For example, in addition to further inquiry regarding the findings here, Edwards' (1993) finding that female students produced significantly more complex concept maps than male students should be looked into in the context of mathematics.

Confidence With Concept Mapping

This aspect involves students' levels of confidence in their concept mapping ability. Students' responses regarding this aspect indicate that in general they were not entirely confident in their concept mapping abilities, but that given the opportunity, they would be able to construct more effective concept maps. This is consistent with their performance on the post-training and post-practice CMS tests, which indicates that their concept mapping skills improved after a period of practice.

Table 7.9 Frequencies, percentages (in parentheses), means, and SD of responses to *confidence* with concept mapping.

Items	Frequency						Mean	SD
	SD	D	LD	LA	A	SA		
35. I am sure I can construct better concept maps if I become more familiar with how to construct them.	0	1 (2.1)	1 (2.1)	3 (6.3)	23 (47.9)	20 (41.7)	5.25	.84
6. Given more time, I can add more propositions to my concept map.	0	0	4 (8.3)	4 (8.3)	24 (50.0)	16 (33.3)	5.08	.87
28*. I feel lost when trying to construct a concept map.	6 (12.8)	23 (48.9)	8 (17.0)	4 (8.5)	6 (12.5)	0	4.40	1.21
22. I am good at concept mapping.	1 (2.1)	8 (16.7)	11 (22.9)	19 (39.6)	7 (14.6)	2 (4.2)	3.60	1.13
33*. I don't think I can do well on concept mapping.	5 (10.4)	7 (14.6)	9 (18.8)	11 (22.9)	15 (31.3)	1 (2.1)	3.44	1.40
11*. I am not sure whether my concept map is good.	1 (2.1)	7 (14.9)	9 (19.1)	15 (31.9)	12 (25.5)	3 (6.4)	3.17	1.22

Topping the list was Item 35 which had a mean of 5.25. The majority of the students agreed with the statement that they could construct better concept maps with more familiarity and that if given more time on a concept mapping task, they could build more propositions. The two students who indicated disagreement with Item 35 also presented weak academic performance, and their mathematics teacher had observed their lack of motivation for school study. Perhaps they likewise lacked motivation to know more about concept mapping and were thus reluctant to put effort into constructing better concept maps.

More than 90 percent of the students agreed with the statement that given more time they could have added more propositions to their concept maps. As you may recall, they had 30 min to construct a concept map with approximately ten given concepts during the four CM tasks described in Chapter 5. When the time was up, some students still appeared to be pondering additional possible connections. Thirty minutes might not be sufficient time to map ten concepts.

The other four items on this aspect presented large standard deviations, indicating that students' reactions to these items varied. Other than their responses to Item 28, which indicates that most student were moderately confidence in their concept mapping ability, their responses to the other three items, with means of 3.17 to 3.60, suggest that they were moderately confident in their ability to perform concept mapping.

Ease of Construction

This aspect relates to students' perceptions of the ease of constructing concept maps. The results are reported in Table 7.10. Items 9 and 31 concern ease of concept mapping in general; Items 19 and 15 are more specific, targeting hierarchy and linking phrase issues in concept mapping. Item 34, a reverse-scored item concerning whether concept mapping is perceived as challenging, may not fit neatly into the construct of *ease*. These observations may explain the low internal consistency calculated for this aspect (Cronbach's α = .339; Cronbach's α = .455 if Item 34 is deleted).

A mean of 4.23, was calculated for Item 9, indicating that the students generally agreed with the statement that a concept map is easy to construct. However, as indicated by the frequencies for Items 31 and 34, over 85 percent of students agreed with the statement that they had to think hard when concept mapping and admitted that they found concept mapping challenging. Students' responses to these items appear to be somewhat contradictory. Such findings are similar to those of Wang (2005). In her study on university students' attitudes toward concept mapping in biology, participants disagreed with the statement that they did not know how to construct concept maps; however, they did not agree that they were good at concept mapping. The students may have found it easy to master concept mapping skills, as indicated by their performance in the CMS tests, but still needed to think hard to construct a concept map of high quality.

In terms of the main steps in concrete mapping steps, determining the hierarchy of the concepts (Item 19) and including accurate linking phrases (Item 15), the students seemed to perceive drafting accurate linking phrases as more difficult

Table 7.10 Frequencies, percentages (in parentheses), means, and SD of responses to *ease of constructing* concept map.

Items	Frequency						Mean	SD
	SD	D	LD	LA	A	SA		
9. Concept mapping is easy to construct.	0	1 (2.1)	9 (19.1)	20 (42.6)	12 (25.5)	5 (10.6)	4.23	.96
19. I find it easy to distinguish the hierarchy among a list of given concepts.	0	5 (10.4)	10 (20.8)	14 (29.2)	13 (27.1)	6 (12.5)	4.10	1.19
15*. I find it difficult to add accurate linking phrases.	3 (6.3)	17 (35.4)	6 (12.5)	15 (31.3)	5 (10.4)	2 (4.2)	3.83	1.31
31*. I have to think hard when doing concept mapping.	0	5 (10.4)	6 (12.5)	10 (20.8)	19 (39.6)	8 (16.7)	2.60	1.22
34*. I find concept mapping challenging.	0	3 (6.3)	6 (12.5)	9 (18.8)	18 (37.5)	12 (25.0)	2.37	1.18

than determining the hierarchy among a list of given concepts. The distribution of the students' responses to Item 15 indicates that they held differing views on the difficulty of including accurate linking phrases, although the mean indicates moderate agreement with the statement that doing so is difficult.

These findings demonstrate that the students had generally favourable attitudes toward concept mapping. They were generally confident in their ability to understand the information embedded in a given map. In general, they reported that concept mapping is easy but challenging and requires hard thinking. They were generally not entirely confident in the quality of the concept maps they had constructed. At the same time, they expressed moderate disagreement with the statement that they felt lost when trying to construct a concept map. They reported experiencing moderate to high levels of enjoyment of during concept mapping and indicated their preference for using it in their further studies. Most students held a positive attitude toward concept mapping and found it a useful tool. Almost all students reported that concept map could reflect their understanding of concepts. Such findings should encourage further exploration and development of concept mappings as a technique for assessing conceptual understanding in mathematics.

Relationship Among the Six Aspects of Students' Attitudes Toward Concept Mapping

This section investigates the relationships among the six aspects of students' attitudes toward concept mapping. Table 7.11 reports the results of the Pearson product-moment correlation analyses. The size of the correlation coefficients is considered an indicator of the strength of a linear relationship between two aspects, and its sign indicates the direction of that relationship.

All aspects are positively related, and there appears to be some general trend in the data. Making inferences, confidence, and ease of constructing a concept map are closely related, with correlation coefficients from .630 to .688, all significant at the 0.001 level (two-tailed). This suggests that a higher perceived ability to make inferences from given concept maps tended to be associated with higher levels of confidence and perceived ease. The three other target aspects,

Table 7.11 Correlational analyses of the relationships between aspects of students' attitudes toward concept mapping.

	Inferences	Ease	Confidence	Enjoyment	Usefulness
Ease	.630*				
Confidence	.688*	.662*			
Enjoyment	.535*	.451*	.534*		
Usefulness	.512*	.220*	.318*	.789*	
Preference	.557*	.236*	.425*	.748*	.784*

$^*p < .001$

enjoyment, usefulness, and preference are also highly correlated (r ranged from .748 to .789, $p < .001$), indicating that higher levels of enjoyment and usefulness are associated with a higher level of preference for using concept maps in the future. Comparatively, the other correlations in the table are either weaker (r ranged from .220 to .557) or not significant. The correlations between ease and usefulness and between ease and preference are less than 0.3, suggesting that students' perception of the usefulness of concept maps and their preference for using concept maps for further study does not rest on the relative ease of constructing a concept map.

Factor analysis indicates that the six aspects of the students' attitudes toward concept mapping can be described by two factors (see Table 7.12).

Table 7.12 presents the rotated factor loadings of absolute values greater than 0.3 only. The first factor includes making inferences, ease of construction, and confidence. I labelled this the cognitive dimension because the three aspects appear related more to the students' cognitive abilities in concept mapping. The second factor covers enjoyment, usefulness, and preference. These three aspects are related to students' affect in concept mapping and so labelled the affective dimension. With the factor analysis, the high correlations between the aspects in Table 7.12 are interpretable because affective beliefs tend to be associated with other affective beliefs and cognitive beliefs with other cognitive beliefs (Trafunow and Sheeran, 1998).

This cognitive-affective difference is also observed by considering the relationships between the students' performance in the four CM tasks and their attitudes toward concept mapping. Table 7.13 shows the result of the correlation analysis between the attitudes and the density and proposition scores of the student-constructed concept maps.

The correlation coefficients indicate that the relationships between the students' concept map scores and their attitudes differed by the aspects of attitudes. The *cognitive* aspects of the students' attitudes toward concept mapping had moderate and statistically significant correlations with their concept mapping scores, while

Table 7.12 VARIMAX rotated factor loadings of the six aspects of students' attitudes toward concept mapping.

Aspect	Factor	
	Cognitive aspects	*Affective aspects*
Inferences	.756	.436
Ease	.905	
Confidence	.856	
Enjoyment	.382	.826
Usefulness		.935
Preference		.900
% of variance	38.81	43.78

Note: Only loadings numerically greater than 0.3 are included

Table 7.13 Spearman's correlations between the concept map scores and attitudes toward concept mapping.

		Cognitive Aspects			Affective Aspects		
		Inferences	Ease	Confidence	Enjoyment	Usefulness	Preference
Proposition score	Algebraic expression	0.542**	0.533**	0.574**	0.258*	0.062	0.169
	Equation	0.524**	0.419**	0.580**	0.214*	0.105	0.163
	Triangle	0.493**	0.390**	0.552**	0.206*	0.094	0.211
	Quadrilateral	0.582**	0.377**	0.615**	0.190*	0.122	0.226
Density	Algebraic expression	0.462**	0.404**	0.676**	0.319*	0.122	0.293*
	Equation	0.426**	0.379**	0.623**	0.301*	0.215	0.274
	Triangle	0.487**	0.406**	0.527**	0.204*	0.076	0.203
	Quadrilateral	0.511**	0.331*	0.540**	0.160*	0.132	0.230

Note: Item 31 and 34 were deleted for a higher Cronbach's α of the scale *ease of constructing CM*
** Correlation is significant at the .01 level (2-tailed)
* Correlation is significant at the .05 level (2-tailed)

the *affective* aspects of the attitudes had only weak correlations with the concept map scores. Such finding is generally consistent among the four topics.

Teachers' Attitudes Toward Concept Mapping

The mean scores and standard deviations of the three aspects of the teachers' attitudes toward concept mapping, i.e. interest in concept mapping, appreciation of the usefulness of concept mapping, and willingness to use concept mapping, are reported in Table 7.14. Since a 5, 4, 3, 2, 1 scoring mode was adopted, the mean score of 3.0 was taken as a benchmark above which the participants are said to be in favour of concept mapping and below which the participants are categorised as not in favour.

The results reported in Table 7.14 indicate that the participants had generally positive attitudes toward concept mapping. They were interested in it and mostly agreed with the statement that concept mapping is useful. They also indicated their willingness to use concept mapping in the future. The in-service teachers showed slightly higher interest in the concept map and stronger willingness to use it than the prospective teachers, although the differences were not statistically significant, as shown by one-way ANOVA. A possible reason for the differences in interest levels might be that the in-service teachers in this study were those who took training courses during summer vacations and had already displayed their motivation toward continuing learning for improving their teaching, perhaps accounting for their greater willingness to accept concept mapping compared to the other participants. Moreover, the in-service teachers had more teaching experience than the prospective teachers and may have had a better idea of how concept mapping could contribute to mathematics education.

Table 7.14 Mean scores and standard deviations of three aspects of attitudes toward concept map.

Aspects	All participants (n=391)	Prospective teachers (university students) (n=173)	Prospective teachers (master students) (n=163)	In-service teachers (n=55)
Interest on concept map	3.80 (0.43)	3.75 (0.45)	3.83 (0.42)	3.85 (0.38)
Appreciation of the usefulness of concept map	3.95 (0.43)	3.98 (0.44)	3.91 (0.44)	3.97 (0.38)
Willingness to use concept map	3.76 (0.39)	3.75 (0.42)	3.75 (0.38)	3.80 (0.32)

Source: Adapted from Jin, Lu, and Zhong, 2015, p. 603

The teachers' responses to open-ended interview questions and essay questions are analysed together because the same questions were used, and the findings are similar. The participants were encouraged to express their ideas in accordance with, but not restricted to, the questions. For example, for the question 'Would you use concept map in teaching?' I told them that it would be valuable if they could identify factors that would motivate their inclusion or exclusion of concept mapping in teaching. Their responses to the questions supplemented their responses to the questionnaire, as described in the following.

First, both the prospective and in-service teachers indicated that the 10-minute introduction I provided did not give them a substantial idea of how concept mapping could be applied in practice. They expressed a desire for further training on concept mapping, including model lessons and concrete examples of its applications. They indicated that they would welcome concentrated training and reference materials which could help them learn more about concept mapping.

Second, they viewed concept mapping as more feasible as a teaching method than as a learning strategy or as an assessment technique. They indicated that it could be used in the classroom to clarify relationships among concepts, i.e. as a model for effectively organising knowledge. At the same time, they opined that concept maps might be useful for teaching some, but definitely not all, mathematics topics. They were not yet sure which topics would be most compatible with concept mapping.

Third, the participants offered the following four factors that could prevent them and, presumably, other teachers from adopting concept mapping for mathematics education.

1   Access. They had limited access to concept mapping through the school systems. Few teachers used concept mapping in school. They had no models. They preferred to rely on traditional teaching methods rather than experimenting with this new technique. This is consistent with their claims that

they needed model lessons or concrete examples to help them get started with concept mapping.

2   Lack of research-based evidence. There is not sufficient research-based evidence in the Chinese context to persuade them that concept mapping is useful for Chinese students. Although there are experimental studies supporting the use of concept maps for mathematics education, such studies are mostly conducted in Western countries. Teachers were not sure whether concept mapping would be suitable for the situation in China.

3   Current evaluation system. The evaluation systems in China are results-focused rather than processes-focused. The strength of the concept map is its ability to illustrate the process of building connections; on the other hand, constructing a concept map is not a feasible task type for large-scale examinations, e.g. university entrance examinations in China. As such, teachers could not really see how concept mapping would contribute to their students' performance on examinations.

4   Tight curriculum. The strength of concept mapping is in the students' active involvement in construction. Directly presenting a teacher-constructed concept map may not contribute much to students' learning. However, it would be time-consuming for students to construct concept maps individually or cooperatively in class. The tight curriculum does not allow much time for incorporating such activities in classroom teaching.

Fourth, the participants were not sure if they were willing to introduce concept mapping to their students and encourage them to use it for learning mathematics because they themselves were not yet familiar with concept mapping and its applications. They opined that whether students would accept concept mapping would depend primarily on whether they appreciated its usefulness at the initial stage. According to this wisdom, care should be taken when introducing concept maps to students, especially in class.

In general, the participating teachers expressed positive attitudes toward using concept mapping in mathematics education. This is consistent with findings by Okebukola (1992) and Wang (2005). Okebukola (1992) investigated Australian teachers' attitudes toward the concept map and the Vee diagram as meta-learning tools in science and mathematics. After a five-day workshop on strategies for improving teacher effectiveness, the teachers responded to a questionnaire wherein they indicated favourable attitudes toward concept mapping in terms of its benefits in facilitating meaningful learning and reducing student anxiety levels. Wang (2005) examined the attitudes toward concept maps among students majoring in chemistry at a normal university in China. For one semester, each participating student constructed a concept map after classroom coverage of each chapter in their physics text. In the present study, the teachers surveyed had relatively less experience with concept mapping. Even so, with only a brief introduction, they too appreciated concept mapping as a potentially useful technique for teaching, learning, and assessment of mathematics. The prospective teachers seemed to be more positive than the in-service teachers on exploring concept mapping in mathematics education. The former had not formally worked as a teacher in schools. It might be easier for them to accept concept mapping because they are not bound by traditional ways of teaching, curriculum schedules, and exam pressures. Researchers wishing to further explore concept mapping for use in mathematics education should consider beginning with prospective teachers because they may have less constraints and more time to digest the information.

Summary

This chapter presents the results of investigating students and teachers' general attitudes toward using concept mapping in mathematics education. In general, the students expressed strong agreement that concept mapping is useful, moderate to high levels of agreement that it is enjoyable, and preference for using the concept map as a tool in their further studies. They were also favourable toward teachers' using concept mapping in mathematics classrooms. These are encouraging findings suggesting that further popularisation of concept mapping in mathematics education is possible.

For teachers, usefulness is necessary but not a sufficient criterion for adopting concept mapping in their classroom teaching; they must also take feasibility and other real issues into account. The teachers were generally interested in concept mapping, expressed positive attitudes toward its uses, and even indicated, in the questionnaire, a desire to apply it in the future. That said, during the informal interview and in response to open-ended questions, teachers indicated their hesitation in actually using the concept map. The main factors driving their hesitation were lack of confidence in the feasibility of concept mapping being used appropriately in class and their doubts as to whether concept mapping can contribute to students' exam performance. Additional operational training and persuasive evidences are needed to convince teachers of the effectiveness of concept mapping and direct their practice.

Note

The teachers' attitudes toward concept mapping has been published as a book chapter: Jin, H., Lu, J., and Zhong, Z. (2015) 'Exploration into Chinese mathematics teachers' perceptions of concept map', in L. Fan, N. Y. Wong, J. Cai and S. Li (eds) *How Chinese teach mathematics: perspectives from insiders*. River Edge, NJ: World Scientific Publishing Co., pp. 591–618.

8   Summary and Implications

Main Findings

This book primarily investigates the feasibility of concept mapping as a technique for assessing students' conceptual understanding in mathematics. In Chapters 2 and 3, I review and analyse the literature and accompanying theory in which the original research described in this book is grounded. Chapter 2 is focused on conceptual understanding, why it is important, how it is defined, and how it is measured. In Chapter 3, I continue reviewing the literature, examine rationales for concept mapping as a method for assessment of conceptual understanding, and thoroughly overview concept mapping, including definitions, applications, formats, and modes of interpretation. Chapter 4 is a guide to concept mapping training in mathematics. It begins with the main training issues involved in concept mapping and compares the effectiveness of different training methods through review of the relevant literature. Chapter 4 also presents the evolution of the training method applied in the main study to prepare the participants for concept mapping, including a description of the preliminary study (for development of a basic programme for mathematics) and pilot study (checking the revised training program before use in the main study). In Chapter 5, I presents the main study on concept mapping. Students map four different mathematical topics after having been prepared with the necessary concept mapping skills (Chapter 4) and having taken some more traditional tests on the same four concepts (Chapter 6) to facilitate comparison. I examine their concept mapping performance and present samples of their well- and poorly constructed concept maps for each of the four topics. Attributes of students' conceptual understanding are assessed through analysis of their performance in mapping individual concepts, including the connections they built between concepts in a pair and the hierarchy and organization of concepts they presented in their maps. In Chapter 6, the concurrent validity of concept mapping for assessment of conceptual understanding in mathematics is examined by comparing students' concept mapping performance with their performance on two traditional school tests on the same basic topics. Finally, in Chapter 7, I present discussion and statistics on students and teachers' attitudes toward concept mapping in mathematics, which are key issue impacting further popularisation of concept mapping in educational settings.

DOI: 10.4324/9781003269373-8

Training on Concept Mapping Skills

The training program in the main study is grounded in the training trials in the preliminary and pilot studies. Three basic concept mapping skills for free-style mapping tasks with given concepts, namely, statement-transformation, simple free association, and extended free association, are defined. Statement-transformation involves correctly conveying given ideas in the form of propositions; simple free association involves students describing connections between given concepts in their own words, and extended free association calls for a holistic view of numerous of propositions.

The students' responses in the interviews and the open-ended written task indicate that, among the three mapping skills, extended free association is the most useful but also the most difficult. Simple free association skills are more difficult to acquire than statement-transformation skills because hard thinking about the relationships is required. Extended free association skills are the most challenging to acquire, but students acknowledged that it was these skills that could benefit them the most. This is because, in addition to formulating linking phrases, it encourages them to seek out connections that they might not have previously realized. Accordingly, students can probably learn more from extended free association than statement-transformation and simple-association.

The results of a follow-up post-training CMS test indicate that the secondary students were able to obtain sufficient concept mapping skills after four 45-minute training sessions. Most of them could correctly express their ideas in the form of propositions and unidirectional links. Although they indicated that constructing linking phrases was quite challenging, they also indicated that they knew that they could only fully illustrate their understanding of the relationships between the given concepts through detailed linking phrases. Moreover, concerning the organisation of the propositions, the concept maps students constructed in the extended free association task are generally legible and their mapping sequences easy to follow. Although not required, over half of the students arranged the concepts hierarchically.

A post-practice CMS test, parallel to the post-training CMS test, was administered after the four CM tasks. At that stage, students had some substantial experience with concept mapping. Accordingly, students' concept mapping skills had improved after the period of practice. However, the only statistically significant improvement observed involved the detailed linking phrases and the organisation of extended free association tasks.

The findings in the preliminary study, the pilot study, and the main study together suggest that students need extensive training on concept mapping before it can be used as a method for assessing their conceptual understanding; otherwise, students' differing levels of concept mapping skills may affect the content validity of the assessment.

Concept Mapping as a Technique for Assessment of Conceptual Understanding: Strengths and Drawbacks

Conceptual understanding is an important learning outcome comprising several interrelated components (Al-Mutawah et al., 2019; Kilpatrick et al., 2001; Pirie and Kieren, 1994). This study evaluates three components of students' conceptual understanding in mathematics, individual concepts, relationships between concepts, and operations of the concepts and their relationships, using the CM tasks, the DEN tasks, and the P&P tasks. The three task types involve a set of common concepts and were administered in Chinese.

The CM tasks in this study are free-style mapping task with a given list of concepts on a mathematical topic. The students' concept maps are analysed by considering the number and correctness of the incoming and outgoing links, and these values are applied to derive insights about students' conceptual knowledge at the level of individual concepts and pairs of concepts and at the whole-class level using collective maps of their performance. These offer glimpses into the cognitive structure shaping students' understanding of the given concepts. Students in this sample were not explicitly instructed to structure the given concepts hierarchically in their concept maps in case they were not mathematically mature enough to do so; however, the collective maps suggest that student concept maps can be analysed to delineate a hierarchy of the given concepts in an embryonic form which parallels the logical structure of the part-whole and subset-set relationships. Data on how students organise concepts is not easily capturable by other measures. Construction of a collective concept map based on individual students' concept maps is a novel mode of analysis inspired by Social Network Analysis. Accordingly, this study not only describes innovate methodologies for accessing deeper knowledge of students' conceptual understanding but also innovative methodologies for studying constructed concept maps.

Comparison of students' performances on the CM tasks and the two traditional school tasks, the DEN- and P&P tasks, revealed additional strengths and limitations of the three techniques in terms of assessing students' conceptual understanding. Among them, they capture both different and related components of conceptual understanding. This is demonstrated by examining specific concepts measured by the three techniques and the correlations among their overall scores. Among the three, the DEN task most directly solicits students' knowledge of the definitions, examples, and nonexamples of individual concepts. However because students select rather than draft the correct answers, they can easily identify the typical examples and nonexamples they are familiar with (Nelson and Pan, 1995) and avoid those that they are unsure about. As such, in assessing knowledge of the definitions, examples, and nonexamples of a concept the DEN task yields useful but limited information about students' understanding of the concept. As Novak and Gowin (1984) argued, the CM task is effective for unravelling the relationships between concepts in pairs and the organizational forms these take within students' cognitive structures. Moreover, the CM task has the power to reveal the centrality of a concept within a specified domain of knowledge, which is consistent with the findings of Schmittau and Vagliardo (2009). However, both the DEN- and CM tasks present limited efficacy in assessing the properties and operations of the concepts. They do not appear to be predictive of students' performance in terms of applying their knowledge of individual concepts and their relationships, in, for example, presenting explanations, solving problems, and building arguments. From this perspective, the P&P task can cover more ground than the DEN- and CM tasks. Teachers can include items addressing definitions, examples, nonexamples, and representations of individual concepts and items asking students to justify relationships between different two concepts or different properties of the same concept. They can also include items involving applications of the relationships and requiring illustration of the rationale of the applications. However, it is limited in that some students may be able to correctly answer the items, especially routine ones, through practice, without a substantial understanding of the relevant concepts (de Vries et al., 2002; Niemi, 1996a). In addition, it can include only a limited number of items, most of which are designed by teachers; compared with the CM task, it is limited in gathering information that students generate spontaneously. It would be difficult to design a P&P task that is as open to capturing direct information about relationships between concepts and the structure of students' knowledge as the CM task.

Concept mapping is unique in teasing out information on students understanding of cognitive links among concepts which cannot be readily ascertained using traditional measures such as the DEN- and P&P tasks administered in this study. Concept mapping is superior to the other tasks in that it focuses on the relationships between concepts and provides information on how students organise the concepts amongst each other, reflecting their cognitive structures. It is limited in its ability to handle the operation issues in conceptual understanding. Moreover, the students did not generally include connections that they were not sure of or for which they found it difficult to draft linking phrases. When prompts were given near completion of the CM tasks, most students added new connections to their concept maps, which suggests that they knew more about the relationships than they had initially demonstrated. This confirms that even when students have had extensive training in concept mapping, their concept maps may not capture everything they know about the concepts. One major implication of these findings is that applying multiple assessment measure and comparing them can lead teachers to more comprehensive views of students' concepts understanding in mathematics.

Topical Differences

Differences were found between students' concept maps on algebraic topics and those on geometric topics. The students constructed fewer links between algebraic concepts than between geometric concepts. On one hand, mathematically speaking, more connections between geometric concepts may in fact exist than connections between algebraic concepts. On the other hand, as indicated by the students in the interviews and in the open-ended written task, they were more familiar with geometric concepts than algebraic concepts. They had recently studied the geometric concepts in the CM tasks and had studied the algebraic concepts some time before. They also reported using geometric concepts more often than algebraic concepts in solving problems. This may account for the greater number of links constructed between the geometric concepts.

Differences between the linking phrases for algebraic topics and those for geometric topics were also noted. For the algebraic topics, most of the linking phrases were definition-based, while the concept maps of the geometric topics included both definition-based and property-based linking phrases. At the secondary school level, geometry is usually taught in terms of properties of shapes and objects; the differences and similarities between the properties of the geometric concepts could enrich these connections. Students are also frequently exposed to the definitions and properties of geometric concepts and their relationships in practice. In contrast, school level algebra is usually taught in terms of manipulations and calculations which emphasise procedural knowledge more than declarative knowledge. This may be one reason the students found it easier to construct concept maps with geometric concepts than with algebraic concepts.

In addition, geometric topics lend themselves more easily to a hierarchical structure than algebraic topics. One reason is that the hierarchical relationships (i.e. set-subset and whole-part) of the geometric concepts are relatively clearer than with algebraic concepts. This finding suggests that concept maps are more suited to gauging students' knowledge of some mathematical concepts than others, depending on how the knowledge is structured within the mathematical subdomains of a school curriculum. If such a hypothesis stands, it might be more suitable to use concept mapping for the teaching and learning of those mathematical concepts than others.

Findings from the correlational analyses and factor analyses of students' scores for the three task-types also confirms topical differences. The task scores for the two algebraic topics yielded a single factor, indicating that the three tasks-types measure the same underlying construct. The task scores for the two geometric topics, however, yielded two different factors, showing that the tasks may focus on different components of conceptual understanding (see Chapter 6 for more details). The DEN tasks for the four topics are parallel because they use the same form to assess students' knowledge of definitions, examples, and nonexamples of the given concepts. The students' different performance in the CM tasks may be the main reason for the different findings in the factor analyses. It should also be noted that the design of the P&P tasks may be another possible reason for the different results obtained. The P&P tasks for the two geometric topics and those of the two algebraic topics were not designed under a unified criterion, although they focused on conceptual understanding in general. The influence of the latter may be eliminated if unified criteria are used to design the P&P tasks in further studies.

Attitudes Toward Concept Mapping

Both students and teachers were generally positive about the use of concept mapping in mathematics. Students were confident in their ability to read information from given concept maps but less confident in their ability to construct concept maps themselves. Most agreed with the statement that concept mapping is easy and at the same time indicated that concept mapping is challenging and requires hard thinking. They indicated moderate to high levels of enjoyment with concept mapping and a moderate preference for using concept mapping in their further study of mathematics. Among the six aspects of attitudes toward concept mapping, the students expressed a significantly positive view of the usefulness of concept mapping. Almost all of the students maintained that a concept map could reflect their understanding of the concepts and considered it fair to judge their conceptual achievement according to their performance in CM tasks. Even with limited knowledge of concept maps, prospective and in-service mathematics teachers seemed to have some general ideas of how to incorporate concept mapping into their classroom teaching (Jin et al., 2015). In addition to basic knowledge on concept mapping to prepare them technologically, their further introduction to concept mapping should include concrete examples of implementing the technique in different stages of classroom teaching and integrating it with the mathematics curriculum.

These findings are encouraging in the context of further exploration of concept mapping and further developing it as a technique for assessment of conceptual understanding in mathematics.

Limitations

There Are Several Limitations to This Study

First, primary data was gathered from a small, and in some sense, specialised, sample, a class of 48 Grade 8 students in China. It is not justifiable to generalise the results to other secondary students without replication of the study with students from diverse backgrounds and grade levels, even within the same country.

Second, for free-style mapping with the given concepts, the 30-minute time limit for the CM task may not have been sufficient time for the students to complete their concept maps. Moreover, the concepts in the given list were not randomly ordered but rather began with the most inclusive concept, followed by more specific concepts. For example, the triangle concepts were ordered triangle, acute-angled triangle, right-angled triangle, obtuse-angled triangle, scalene triangle, isosceles triangle, equilateral triangle, axis of symmetry, angle, median, and midline. Such a given order may impose limits on students' concept mapping such as by discouraging them from displaying their understanding of the concepts in their own manner. Moreover, the prompts provided near the end of the students' concept mapping task may have influenced, in particular, the connection strengths illustrated in the collective maps. These issues present limitations to interpreting the findings from the students' concept maps.

Third, concept maps can be coded and scored in multiple ways, for example, by holistic scoring or structural scoring methods (McClure et al., 1999). In addition to awarding points for correct propositions, such methods allow for consideration of students' overall understanding of given concepts as represented by a concept map and of the overall hierarchical structure of the map as a whole. If different scoring methods had been applied to assess the student-constructed maps, the results of the comparative analysis the CM-, DEN-, and P&P task scores may have been different. Multiple scoring methods should be applied to obtain more robust understandings of the relationships among these variables.

Fourth, the design of the P&P tasks needs to be refined. The P&P tasks designed for this study follow traditional school test models and thereby focuses more on students' understanding of concepts rather than their ability to apply them in solving problems. The correlational analyses of the relationships among the tasks indicate that it would be worthwhile to classify the items in the P&P tasks into individual concepts, relationships among individual concepts, and relationships between concepts and operations, corresponding to the three components of conceptual understanding. In this study, the categories are not considered in the correlational analyses since the number of items for a category are not balanced across the tasks. Additional information may be available if the correlations are examined in consideration of the three categories. Moreover, it should be noted that the P&P tasks did not cover the same exact concepts as the corresponding CM- and DEN tasks, though overall they were on the same topics. This may have affected the correlations of the P&P task scores with the scores of the other two types of tasks.

Implications for Teaching, Learning, and Assessment

The findings presented in this book have multiple, meaningful implications for teaching, learning, and assessment in schools.

First, the CM-, DEN-, and P&P- tasks appear to assess the strengths and weaknesses of the students' conceptual understanding of the four mathematical topics from different perspectives. For example, the DEN tasks revealed that the students were not familiar with the definitions of some algebraic concepts, e.g. degree and fractional expression. The P&P tasks revealed that the students were better at recognising routine examples and nonexamples than non-routine ones, an indication that they did not fully understand the concepts. The CM task results indicate that the students were not skilled at making connections between coordinate concepts. These research-based findings suggest that the results of the CM tasks can reveal meaningful information about students' conceptual understanding that can inform their instructional decisions for each mathematics topics at the individual student and whole-class level.

Second, the present study has implications in terms of what it would take for school teachers to adopt concept mapping for assessing students' conceptual understanding. As noted, Chapter 4 presents a detailed guide to preparing students with the concept mapping skills necessary to complete mapping tasks. Absent an intensive training program on concept mapping such as was conducted for this study, which is unlikely in a school setting, teachers wishing to adopt it for assessment purposes should first allow sufficient time for their students to gain familiarity with concept mapping. It is suggested that school teachers and students start concept mapping with a small number of concepts, such as three to six (see the CMS tests in Chapter 4), because the more concepts involved, the more challenging concept mapping can be, especially for those new to the process. The definition of concept map used in this book is rather strict. According to this definition, nodes must represent concepts rather than phrases, and such nodes must be linked by unidirectional arrows with accompanying detailed linking phrases to form propositions. Such restrictions are suitable for creating concept maps from a list of given concepts and makes scoring relatively easy. Teachers may wish to relax some of these constraints. For example, the nodes of a concept map could be phrases, theorems, or statements, and bidirectional links could be permitted. The need for such adaptations will likely arise naturally as teachers gain experience and flexibility in using concept mapping in classroom teaching.

Third, other than being used solely for evaluating and grading students' learning outcomes, concept mapping can be an integral part of the learning process itself. Researchers who advocate for the use of concept mapping argue that the process encourages students to review and reflect on what they have learned, to examine and explore interrelations among concepts, and even revise their understandings (e.g. Baroody and Bartels, 2000). That being said, initial training on concept mapping is time-consuming, and concept mapping skills seem to be cumulative, increasing over time with practice. As such, using concept mapping solely for assessment purposes may not be economical. Rather, it would be more practical and beneficial to first incorporate concept mapping gradually as a learning strategy. In this way, students would gain familiarity with concept mapping and its skills and requirements without the time and effort involved in intensive training sessions.

Fourth, this study contributes to the literature concerning the use of concept maps as an assessment technique for conceptual understanding in mathematics in the following five areas:

1   Students' concept mapping training was successful. The concept mapping skills defined and the training programmes and methods of the evaluating described in this book can be adapted for mathematics teaching, learning, and assessment in other contexts within China and beyond.
2   The construction of collective maps based on individual students' concept maps presents a novel analysis model that yields meaningful information on students' conceptual understanding at the whole-class level. Further research on this technique for other topics may include additional indices of links used in Social Network Analysis, such as centralities and betweenness.
3   Inconsistent findings regarding the validity of concept mapping in education are reported in the literature. Inconsistency among these studies are likely due to differing concept mapping task formats for different topics and differing grade levels among participants. The present study contributes to knowledge on the concurrent validity of free-style concept mapping with a given concept list as an assessment technique.
4   Investigation into students and teachers' attitudes toward concept mapping is encouraging and suggests that further adoption of concept maps for assessment purposes in mathematics is feasible.

5   This study found topical differences regarding the use concept mapping as an assessment technique. These differences may partially explain the inconsistent findings on the validity of concept map-based assessment in the literature (e.g. Rice et al., 1998; Shavelson et al., 2005).

Implications for Further Research

The studies introduced in this book can be extended and enriched in several directions, as follows:

First, although it was found effective for Grade 8 students in China, the training programme detailed in this book is time-consuming, comprising four 45-minute sessions. Further studies developing simplified training programmes for increased efficiency, economy, and feasibility are in order.

Second, this study reveals differences between algebraic topics and geometric topics concerning the concurrent validity of concept mapping as an assessment of conceptual understanding in mathematics. This issue should be further studied because the students' scores on the example and nonexample task for the two geometric topics presented little variation, and no uniform criterion was used in designing the P&P tasks for the algebraic and geometric topics. These limitations may be affecting the concurrent validity results reported in this book.

Third, methods for analysing student-constructed concept maps can be extended, for example by adapting methods from Social Network Analysis to the study of cognitive networks (Jin and Wong, 2010). Moreover, although students' concept maps provide insights into their knowledge structures, this study did emphasise analysis of such structures. Issues regarding the hierarchies presented in the students' concept maps were not directly addressed. Further research exploring this aspect of concept mapping may be needed.

Fourth, the effects of using concept mapping as an assessment of students' conceptual understanding was not examined in this study. Students' responses in the ATCMQ, interview, and open-ended written task evidence that they benefited from concept mapping. The elaboration effect of concept mapping on conceptual understanding in mathematics is worthy of further study.

Fifth, the CM tasks did not elicit students' full understanding of the given concepts and connections. Prompts and interviews may be needed to gather further information about the reasons for missed connections and incomplete linking phrases.

Finally, most of the items in the ATCMQ were designed by the present researcher. The Cronbach's α of one of its aspects (*ease of construction*) indicates relatively weak internal consistency for the items. Further research is required to refine and validate this questionnaire.

Closing Remarks

To conclude, helping students develop conceptual understanding in mathematics is crucial, and it is equally crucial that teachers evaluate their students' conceptual understanding in-depth (Crooks and Alibali, 2014; Nickerson, 1989). Though some researchers have suggested possible assessment styles and forms (e.g. Niemi, 1996a, 1996b; White and Gunstone, 1992), the variety of assessment techniques and accompanying tools (tasks) sufficient for measuring students' conceptual understanding in mathematics have not yet been developed. This book is consistent with current recommendations from education professions in math and science, and it demonstrates how concept mapping is one such important technique. The training programme, the CM tasks, and methods of data analysis are intended for use by other researchers and educators for inclusion with or without modification in their assessment programmes. It is hoped that this study catalyses further investigations into this important area in mathematics education.

Appendices

Appendix A

Concept Mapping Skill (CMS) Test in the Pilot Study

Name: _____ Class: _____ _____

Class Register No.: _____ Gender: _____ Age: _____.

Instructions:

- The purpose of this test is to measure your concept mapping skills.
- This test has NOTHING to do with your school performance and your understanding of the concepts involved.
- Please attempt all of them as best as you can, according to the requirements specified in the training section.

1. Statement Transformation

   (1) A triangle has three median lines.
   (2) Two plane figures are called congruent figures if they can coincide exactly with each other.
   (3) Two lines in a plane are either parallel lines or intersecting lines.
   (4) Two numbers with different symbols but same absolute value is called opposite numbers.

2. Simple Free Association

   (1) Scalene triangle, triangle
   (2) Linear inequality with one unknown, unknown
   (3) Even numbers, odd numbers, prime numbers
   (4) Square root, rational number, irrational number

3. Extended Free Association: construct a concept map using the given concepts.

   Parallel lines, intersecting lines, perpendicular lines, perpendicular bisector, equidistance, middle point

   Note: please check your concept map: (1) whether it includes all the given concepts, (2) whether all the links are directional, (3) whether you have added as many meaningful links as you can, and (4) whether the linking phrases are completed and accurate.

Post-Training CMS Test

Name: _____ Class: _____ _____
Class Register No.: _____ Gender: _____ Age: _____.

Instructions:

- The purpose of this test is to measure your concept mapping skills.
- This test has NOTHING to do with your school performance and your understanding of the concepts involved.
- Please attempt all of them as best as you can, according to the requirements specified in the training section.

1. Statement Transformation

   (1) The sum of several monomials is called polynomial.
   (2) Real numbers are one-to-one correspondent to the points in a number line.
   (3) Two lines in a plane are either parallel lines or intersecting lines.
   (4) Two numbers with different symbols but same absolute value is called opposite numbers.

2. Simple Free association

   (1) Scalene triangle, triangle
   (2) Linear inequality with one unknown, unknown
   (3) Even numbers, odd numbers, prime numbers
   (4) Quadratic radical expression, rational number, irrational number

3. Extended Free association: construct a concept map using the concepts given below.

   Parallel lines, intersecting lines, perpendicular lines, perpendicular bisector, equal distance, middle point

   Note: please check your concept map that (1) whether it includes all the given concepts, (2) whether all the links are arrowed, (3) whether you added as many meaningful links as you can, and (4) whether the linking phrases are as completed and accurate.

Post-Practice CMS Test

Name: _____ Class: _____
Class Register No.: _____ Gender: _____ Age: _____.

Instructions:

- The purpose of this test is to measure your concept mapping skills.
- This test has NOTHING to do with your school performance and your understanding of the concepts involved.
- Please attempt all of them as best as you can, according to the requirements specified in the training section.

1. Statement Transformation

    (1) A triangle has three median lines.
    (2) Irrational numbers cannot be represented in the form of fractions.
    (3) Two plane figures are called congruent figures if they can coincide exactly with each other.
    (4) Inequalities are mathematical sentences which use inequality symbols to express inequality relations.

2. Simple Free association

    (1) Rectangle, parallelogram
    (2) Linear equation with two unknowns, unknown
    (3) Addition, subtraction, multiplication
    (4) Natural numbers · prime number · composite number

3. Extended Free association: construct a concept map using the concepts given below.

    Real number, number line, zero, origin, absolute value, opposite numbers
    Note: please check your concept map that (1) whether it includes all the given concepts, (2) whether all the links are arrowed, (3) whether you added as many meaningful links as you can, and (4) whether the linking phrases are as completed and accurate.

Appendix B

DEN Task & CM Task (Algebraic Expression)

Name: _____

School: _____

Class: _____

Class Register No.: _____

Instruction:

- The purpose of this test is to find out your understanding of concepts related to algebraic expression.
- Please fill in the required information above.
- This test includes two tasks

  (1) *Fill in a table*: it aims to find out what you remember about the concepts, i.e. definitions, examples, and nonexamples;
  (2) *Construct a concept map*: it intends to find out what you know about the relationships between the given concepts.

- Please follow the instructions for each task and answer accordingly.

Thanks for your cooperation！！

Time allowed: 45 minutes (15 minutes for Task 1, and 30 minutes for Task 2)

Concepts Related to Algebraic Expression

A list of concepts is provided in the following table. Provide their definitions, examples, and nonexamples.

No.	Concepts	Definition/ Description	Examples	Nonexamples	Note
1	Integral expression				
2	Monomial				
3	Polynomial				
4	Fractional expression				
5	Coefficient				
6	Degree				
7	Constant term				
8	Like terms				
9	Unlike terms				
10	Common factor				
11					
12					

_____

_____

_____

_____

Class: _ Register No: _____

Concept Mapping : Please construct a concept map using the concepts given in the table above.
(Notes : ① Please try to include as many meaningful connections as you can; ② please try to use complete and accurate linking phrases)

Appendix C

An Example of Students' Responses to DEN Task & CM Task (Triangles)

三角
四边形相关概念

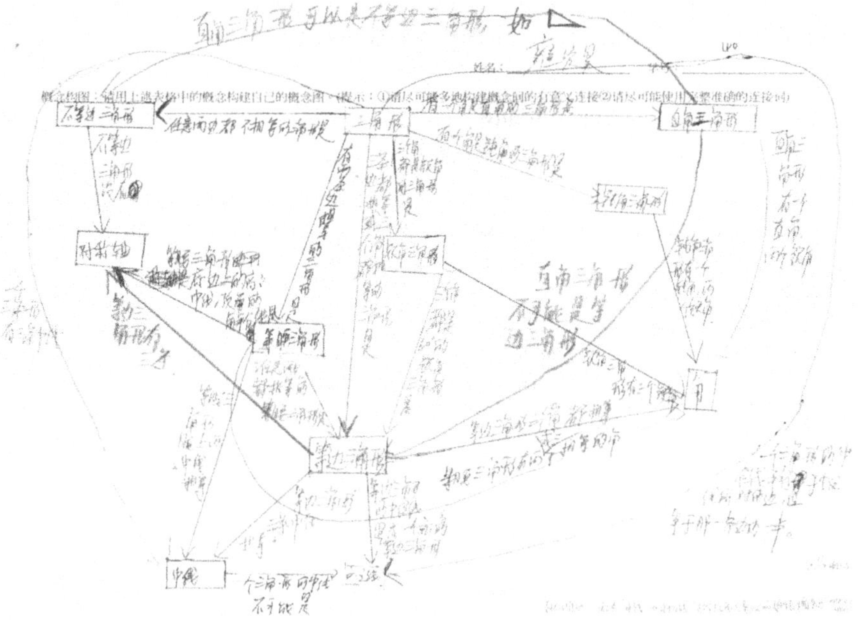

Appendix D

Traditional Paper-and-Pencil Task (P&P Task) (Algebraic Expression)

—Answers and Marking Scheme

1. Which of the following expressions are integral expressions? Which are fractional expressions? Which are both integral and fractional expressions? Which are neither? Please tick the correct box. (1' for each, thus full mark 0.5' × 8 = 4'; the answers are marked with √)

	Integral Expressions	Fractional Expressions	Both	Neither
(1) $-\dfrac{5}{8}a^2$	√			
(2) $\dfrac{3}{5}m = 0$				√
(3) $\dfrac{m-n}{5}$	√			
(4) a	√			
(5) $\dfrac{2}{5}$	√			
(6) $\dfrac{b}{a-2} = 0$				√
(7) $a^2 + b + 7 = 0$	√			
(8) $3ab + \dfrac{2a}{b}$		√		

(If a student did not provide any answer to an expression or ticked two boxes, the response to that expression is considered invalid. Same for Item 2)

2. Which of the following expressions are monomials? Which are polynomials? Which are both monomials and polynomials? And which are neither? Please tick the correct box (1' for each, thus full mark 1' × 8 = 8'; the answers are marked with √)

	Monomials	*Polynomials*	*Both*	*Neither*
(1) $4a + 13$		√		
(2) $-\dfrac{5}{8}a^2$	√			
(3) $\dfrac{m^2 - n}{2}$		√		
(4) b	√			
(5) $\dfrac{3}{a}$				√
(6) $\dfrac{-m^2 n}{5}$	√			
(7) $a^2 + b = 7$				√
(8) $3ab + \dfrac{2a}{b}$				√

3. *a* and *b* are constants. State the coefficient of x^2 in each of the following expressions.

(1) $x^2 + 5x - 6$ <u>1</u>

(2) $-x - 2x^2 - 7y$ -2

(3) $4ax^5 + 2x^3 - 7ax^2 + 2bx - a$ <u>$-7a$</u>

(4) $-(2a)^2 x + (bx)^2 + ab^2 x^2$ <u>$b^2 + ab^2$</u>

Marking scheme: Full mark 4'; 1' for each expression. Correct answers are provided on the lines.

4. State the number of terms and degrees of the following expressions (Full mark: 1' × 4 + 1' × 4 = 8')

(1) $5a - 8b + 2^2$ No. of terms: 3   Degree: 1.

(2) $-3x^2y + 5xy^3 - 7xy + y - 1$ No. of terms: 5   Degree: 4.

(3) $4p(q^2 - 3r)$ No. of terms: 2   Degree: 3.

(4) $\dfrac{1 - r^3}{5}$ No. of terms: 2   Degree: 3.

Marking scheme: Full mark 8'; 2' for each expression, 1' for its degree, 1' for its number of terms. Correct answers are provided on the lines above.

5. *a* and *b* are constants. State the constant terms of the following expressions (Full mark 2').

(1) $a^2 - 2ab^2 + bx$ $\underline{\quad a^2 - 2ab^2 \quad}$

(2) $x^2 - \dfrac{2}{3}x^2y - \dfrac{1}{2} + y^2$ $\underline{\quad -1/2 \quad}$.

Marking scheme: Full mark 2'; 1' for each expression. Correct answers are provided on the lines above.

6. For each of the following expressions, state the common factors among the given terms (Full mark 2').

(1) $18a^2 + 24ab^2$ $\underline{\quad 6a \quad}$

(2) $a^2x^{n+2} + abx^{n+1} - acx^n - ax^{n+3}$ $\underline{\quad axn \quad}$

Marking scheme: Full mark 2'; 1' for each expression. Correct answers are provided on the lines above.

7. Decide whether the following terms are like terms. Show your answers by ticking the corresponding word. (Full mark 1' × 5 = 5')

(1) x^2y and xy^2 True/**False**

(2) $-m^3n^2$ and $3n^2m^3$ **True**/False

(3) $4ab$ and $4a^2b^2$ True/**False**

(4) $-6a^3b^2c$ and cb^2a^3 **True**/False

(5) 4^3 and 5^2 **True**/False

Marking scheme: Full mark 5', 1' for each expression. Answers are bold.

8. Decide whether each of the following statements is True or False. Tick "√" if it is correct; otherwise, tick "×" to show it is wrong (Full mark 1' × 4 = 4').

(1) Like terms have same coefficients; unlike terms have different coefficients.   (×)

(2) Like terms have same degree; unlike terms have different degree.   (×)

(3) Like terms can be added; unlike terms cannot be added.   ($\sqrt{}$)
(4) Like terms can be multiplied; unlike terms cannot be multiplied.   ($\times$)

Marking scheme: Full mark 4', 1' for each expression. Answers are provided above.

9. Decide whether the following statements are True or False. Tick "$\sqrt{}$" if it is correct; otherwise, tick "$\times$" to show it is wrong (Full mark 1' $\times$ 4 = 4').

   (1) A monomial multiplies a polynomial, their result has one more term than the polynomial.   ($\times$)
   (2) An integral expression could include fractions.   ($\sqrt{}$)
   (3) Constant terms are terms without letters.   ($\sqrt{}$)
   (4) Factorisation will not change the number of terms in a polynomial.   ($\sqrt{}$)

   Marking scheme: Full mark 4', 1' for each expression. Answers are provided above.

10. (1) Find the value of x such that after simplification $\frac{2}{3}x - 5y - 5$ becomes a monomial.

   Marking scheme: Full mark 2', 1' for equation '$\frac{2}{3}x = 5$', 1' for result '$x = 15/2$'.

   (2) Find the value of m such that the degree of polynomial $-0.7x^{(-0.5m-4)}y^2 + x^2y - 3$ is 4.

   Marking scheme: Full mark 2', 1' for equation '$-0.5m - 4 = 2$', 1' for result '$m = -12$'.

   (3) Find the value of $m + n$ such that monomial $2x^2y^m$ and $-\frac{1}{3}x^ny^3$ are like terms.

   Marking scheme: Full mark 2', 1' for values '$m = 3$, $n = 2$', 1' for result '$m + n = 5$'.

11. Given that fractional expression $\dfrac{2x - 4}{|x| - 5}$,

   (1) what is the value of x so that this expression is meaningful? (Full mark 2')
   (2) what is the value of x so that the value of the expression equals to 0? (Full mark 2')

   Marking scheme: Full mark 4'.
   (1) $|x| - 5 \neq 0$ ...... 1'
   $x \neq \pm 5$ ...... 1'
   (2) $2x - 4 = 0$ and $x \neq \pm 5$ ...... 1'
   $x = 2.$ ...... 1'

12. Express the following word statements algebraically.

   (1) Subtract $3m$ from the quotient of n divided by p.

   Marking scheme: Full mark 1'.

   (2) Three consecutive even numbers, if $2n$ is the intermediate even number, what are the other even numbers?

   Marking scheme: Full mark 2'. 1' for $2n - 2$, 1' for $2n + 2$.

13. Xiao Ming goes to a movie with his family. He has 2 free tickets. Find the amount he has to pay for the tickets if there are n family members and each ticket costs $\$p$. (Full mark 2')

    Marking scheme: Full mark 2' for correct answer ' $(n-2)p$ '.

14. Given $m = 2n + 3$ and $n = 7$, find the value of m.

    Marking scheme: Full mark 1' for correct answer '$m = 17$'.

15. Given $(x + 1)^3 + x = 349$ when $x = 6$; find the value of x when $(5x + 1)^3 + 5x = 349$. (4')

    Marking scheme: Full mark 3'.

    $$5x = 6. \ldots \ldots 2'$$
    $$x = 6/5. \ldots \ldots 1'$$

16. Xiao Wang says that $3a^2 + 2a + 5a^2 = 10a^3$ and $7b \times 5b^2 \times 6 = 210b$.

    Do you agree with Xiao Wang's equations? If not, what is or what are the mistake(s).
    Marking scheme: Full mark 4'. 2' for reasoning, 2' for answers.

    $$3a^2 + 2a + 5a^2 = 8a^2 + 2a \ \ldots \ldots 1'$$
    $$7b \times 5b^2 \times 6 = 210b^3 \ \ldots \ldots 1'$$

    Reasoning: Unlike terms cannot be added; when multiply, should add the exponents of the same variable. . . . . . 2' (2' for completed reasoning; 1' for partly correct reasoning)

17. Write an algebraic expression that has 3 terms involving the variables p and q.

    Marking scheme: Full mark 2' for correct answer, e.g. $1 + p + q$ or $p + q + 2pq$.

18. Simplify each of the following expressions (Full mark 4'):

    (1) $-3cx + 8x - 1 + cx + 5x - 2$

    Marking scheme: Full mark 2'

    Recognize like terms. . . . . . 1'
    Result: $-2cx + 13x$—3. . . . . . 1'

    (2) $xy - 3x^2y^2 - xy^2 - 3xy - 2xy^2 + x^2y^2 - y$

    Marking scheme: Full mark 2'

    Recognize like terms. . . . . . 1'
    Result: $-2xy - 2x^2y^2 - 3xy^2 - y$ . . . . . . 1'

19. Factorize the following expressions (Full mark 4'):

    (1) $\frac{1}{4}x + x^3 - x^2$

    Marking scheme: Full mark 2'

    Recognize common factor. . . . . . 1'

    Result: $x(x - \frac{1}{2})^2$ . . . . . . 1'

    (2) $a^3 x^{n+2} + 2a^2 bx^{n+1} + ab^2 x^n$

    Marking scheme: Full mark 2'

    Recognize common factor. . . . . . 1'

    Result: $ax^n (ax + b)^2$ . . . . . . 1'

20. Zhang Hua made a mistake in an exam. When computed a polynomial plus $5xy - 3yz + 2xz$, he took the addition as subtraction and got the answer $2xy - 6yz + xz$. Find the correct answer for him.

    Marking scheme: Full mark: 3'.

    Correct expression $2xy - 6yz + xz + 2(5xy - 3yz + 2xz)$ . . . . . . 2'

    Final answer: $12xy - 12yz + 5xz$ . . . . 1'

21. Pose a problem that could be represented by the expression $n + 4(m - 2)$.

    Marking scheme: Full mark: 4'.

    Use m—2 or $4(m$—2) . . . . . . 2'

22. Xiao Li multiplies a number by 5 and then adds 12. She then subtracts the original number and divides the result by 4. She notices that the answer she gets is 3 more than the number she started with. She says, "I think that would happen, whatever number I started with." Is Xiao Li's conclusion correct? Using algebra, show whether she is right.

    Marking scheme: Full mark: 4'

    Decoding $\frac{[(x \cdot 5) + 12] - x}{4}$ . . . . . . 2'

    Conclusion ... $= x + 3$ . . . . . . 2'

Traditional Paper-and-Pencil Task (P&P Task) (Equation)

—Answers and Marking Scheme

1. Which of the following expressions are linear equations with one unknown? Which are linear equations with two unknowns? Which are both linear equation with one unknown or linear equation with two unknowns? Which are neither? Please tick the correct box. (0.5' for each; Full mark 0.5' × 15 = 7.5', the answers are marked with √)

	Linear equation with one unknown	Linear equation with two unknowns	Both	Neither
(1) $2x + y = 10$		√		
(2) $xy = 3$				√
(3) $2x^2 - y = 9$				√
(4) $\dfrac{1}{x-y} = 2$				√
(5) $7 + 3x = 14$	√			
(6) $2x + \dfrac{2}{y} = 1$				√
(7) $x^2 - 2x - 3 = 0$				√
(8) $5 + 15 = 20$				√
(9) $\dfrac{x}{4} + \dfrac{y}{3} = 1$		√		
(10) $x - 3y$				√
(11) $xy = x - y$				√
(12) $s = 5t$		√		
(13) $\dfrac{y-1}{3} = 2y - 3$	√			
(14) $3x - \dfrac{1}{4} = 8$	√			
(15) $\dfrac{1}{2}x = 1$	√			

2. Which of the following functions are linear functions? Which are proportional functions? Tick the correct box. (1' for each; Full mark 1' × 6 = 6', the answers are marked with √)

	Linear function	Proportional function	Both	Neither
(1) $y = -\dfrac{x}{3}$			√	
(2) $y = -\dfrac{8}{x}$				√
(3) $y = 8x^2 + x(1-8x)$			√	
(4) $y = \dfrac{1}{x+1}$				√
(5) $y = 1 + x$	√			
(6) $y = 5x$			√	

3. Which of the following value is the solution of equation $2x + 3 = 9$.　　(B)

    (A) 2　　　　(B) 3　　　　(C) 4　　　　(D) 9

    Marking scheme: Full mark 2' for correct answer 'B'. Other answers, assign 0'.

4. How many sets of positive integral solutions does equation $4x + y - 20 = 0$ have? (A)

    (A) 4　　　　(B) 5　　　　(C) 6　　　　(D) 7

    Marking scheme: Full mark 2' for correct answer 'A'. Other answers, assign 0'.

5. The graph shows linear function $y = ax + b$ and $y = kx$ intersect at point P. Based on the information given in the graph, find the solution of simultaneous linear equation with two unknowns $\begin{cases} y = ax + b \\ y = kx \end{cases}$: $\begin{cases} x = -4 \\ y = -2 \end{cases}$

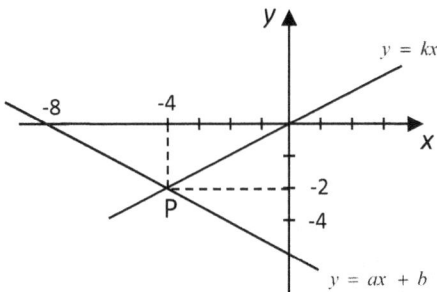

Marking scheme: Full mark: 2', 1' for partly correct answers. Did not write in a correct format, i.e. write $x = -4$, $y = -2$, assign 1'; Get one value correct, assign 1'. Some students wrote the values for k, a, and b. If these three values are all correct, assign 1' because it shows they could read the graph. Other answers assign 0'.

6. In linear function $y = kx - k \left(k \neq 0 \right)$, the value of y decreases while the value of x increases, which of the following graphs represents the function   (D)

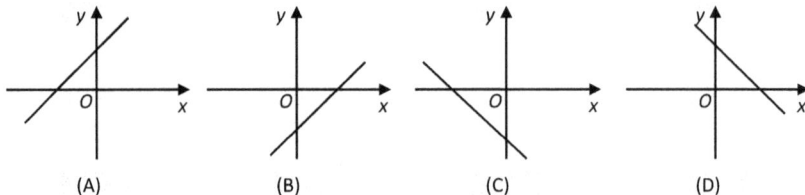

(A)   (B)   (C)   (D)

Marking scheme: Full mark: 2'. Other answers, assign 0'.

7. Sam wanted to find three consecutive even numbers that add up to 84. He wrote the equation $k + (k + 2) + (k + 4) = 84$. What does the letter k represent?

(B)

(A) The average of the three even numbers
(B) The least of the three even numbers
(C) The middle even number
(D) The greatest of the three even numbers

Marking scheme: Full mark: 2'. Other answers, assign 0'.

8. State whether each of the following statements is True or False (Full mark: 4').

(1) If the independent variable and dependent variable in a function are in a direct proportion, the function must be a linear proportional function. (√)
(2) In a linear proportional function, the independent variable and dependent variable are in a direct proportion. (√)
(3) In a linear function, if the coefficient of the independent variable is a positive number, then the function is a linear proportional function. (×)
(4) Proportional function is not necessarily a linear function. (×)

Marking scheme: Full mark 4', 1' for each. Answers are provided above.

9. State whether each of the following statements is True or False. (Full mark: 4')

(1) When an expression has an equal sign, it must be an equation. (×)
(2) Equation must have an equal sign. (√)
(3) If an equation has two unknowns and the exponent of each unknown is 1, the equation is called linear equation with two unknowns. (×)
(4) The coefficient of the unknown in a linear equation in one unknown is not necessarily to be 1. (√)

(5) The values of two unknowns that satisfy any one of the equations in simultaneous linear equations with two unknowns is called the solution of the simultaneous equations. (×)

(6) For a linear function $y = kx + b(k \neq 0)$, if we know how the value of y changes while the value of x increases, we can infer whether the value of k is larger or smaller than zero. (√)

Marking scheme: Full mark 6', 1' for each. Answers are provided above.

10. State whether each of the following statements is True or False (Full mark: 4').

(1) The abscissa of the intersection of $y = kx + b(k \neq 0)$ and the x-axis is the solution of the linear equation $kx + b = 0$. (√)

(2) The coordinate of any point on the graph of $y = kx + b(k \neq 0)$ is the solution of linear equation $kx - y + b = 0$. (√)

(3) The point (x, y), where x and y are the solutions of $kx - y + b = 0$ is not necessarily a point on the graph of $y = kx + b$ $(k \neq 0)$. (×)

(4) The values of x, which makes the function $y = kx + b(k \neq 0)$ larger than 0, are the solutions of linear inequality $kx + b > 0$. (√)

Marking scheme: Full mark 4', 1' for each. Answers are provided above.

11. If the graph of a proportional function passes through point (2, 1), then the equation of the function is $y = x/2$.

Marking scheme: Full mark 2' for correct answer $y = x/2$. Other answers, assign 0'.

12. If $x = 4$ is the solution of the equation $mx - 8 = 20$, then $m = 7$.

Marking scheme: Full mark 2' for correct answer $m = 7$.

13. $x^{|m|-3} + (m - 4)y + 2 = 0$ is a linear equation with two unknowns x and y, find the value of m.

Marking scheme: Full mark 3'.
$$|m| - 3 = 1 \ldots \ldots 1'$$
$$m - 4 \neq 0 \ldots \ldots 1'$$
$$m = -4 \ldots \ldots 1'$$

14. Give an equation of the form $\dfrac{x - a}{b} = c$, where a, b and c are constants, such that the solution of the equation is $x = 4$.

Marking scheme: Full mark 2'.

15. Given function $y = (k + 1)x + k^2 - 1$, (1) what is the value of k when it is a linear function? (2) what is the value of k when it is a proportional function? (Full mark 4')

Marking scheme: Full mark 4'.
(1) $k + 1 \neq 0 \ldots \ldots 1'$ (Category II step)
$k \neq -1 \ldots \ldots 1'$ (Category III step)
(2) $k + 1 \neq 0$ and $k^2 - 1 = 0 \ldots \ldots 1'$ (Category II step)
$k \neq 1 \ldots \ldots 1'$ (Category III step)

16. If linear function $y = (m+4)x - 3 + n$ (where x is the independent variable) intersects the y-axis at a negative value, find the respective range for m and n. (Full mark: 4')

   Marking scheme: Full mark 4'.
   Linear function => $m + 4 \neq 0$. . . . . . 1' (Category II step)
   $\qquad\qquad\qquad m \neq -4$. . . . . . 1' (Category III step)
   Intersects y-axis at a negative value => $-3 + n < 0$. . . . . . 1' (Category II step)
   $\qquad\qquad\qquad\qquad\qquad\qquad\qquad n < 3$. . . . . . 1' (Category III step)

17. A company needs to produce 120 disks. After worker A worked on it for 1 hour, worker B joined. They finished the assignments after they worked together for 3 hours. Given that B makes 5 disks less than A per hour, find how many disks do worker A and B make per hour respectively.

   Marking scheme: Full mark 4'.
   Assume A can make x disks per hour, then B can make $x - 5$ disks. . . . . . 1'
   Thus $\qquad\qquad\qquad\qquad x + 3[x + (x - 5)] = 120$ . . . . . . 1'
   $\qquad\qquad\qquad\qquad\qquad x = 135/7$. . . . . . 1'
   $\qquad\qquad\qquad\qquad\qquad x - 5 = 100/7$. . . 1'

18. Look at the numbers in this table and answer the questions. (Full mark: 6')

x	0	1	2	3	4	5	6
y	2	5	8	11	14	17	. . .

   Marking scheme: Full mark 6'.
   (1) When x is 5, what is y?
   $\qquad\qquad\qquad 17$. . . . . . 1'
   (2) When x is 6, what is y?

   $\qquad\qquad\qquad 20$. . . . . . 1'

   (3) When x is 60, what is y?
   $\qquad\qquad\qquad 182$. . . . . . 2'
   (4) Use algebra to write a rule connecting x and y:
   $\qquad\qquad\qquad y = 3x + 2$

Traditional Paper-and-Pencil Task (P&P Task) (Triangle)

—Answers and Marking Scheme

1. Which of the following figures are triangles? Please write the corresponding letters on the line below (Full mark: 2').

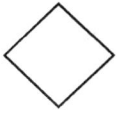   

A B C D

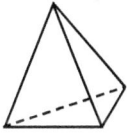  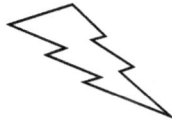 

E F G H

Triangle: *D F H*   .

Marking scheme: Full mark 2'; 1' for partly correct answer. The correct answer is provided on the line above.

2. State whether the following statements are true for EVERY triangle. Show your answers by ticking the corresponding word. If you cannot decide, tick "Uncertain" (Full mark: 1' × 6 = 6').

   (1) It has three sides. **Yes**/No/Uncertain
   (2) The sum of all its angles is 180°. **Yes**/No/Uncertain
   (3) All its sides are equal in length. Yes/**No**/Uncertain
   (4) All its sides are *not* equal in length Yes/**No**/Uncertain
   (5) It has at least one symmetrical axis. Yes/**No**/Uncertain
   (6) It has at least one acute angle. **Yes**/No/Uncertain

   Marking scheme: Full mark 1' for each; if choose wrongly or choose 'uncertain', assign 0'. Answers are bold, hereinafter the same.

3. A triangle has two sides of equal length and two of its angles of equal size. Decide whether the following statements about this triangle are True or False (Full mark: 1' × 5 = 5').

   (1) The triangle cannot be equilateral. True/**False**/Uncertain
   (2) The triangle is an isosceles triangle. **True**/False/Uncertain
   (3) The triangle cannot be a right-angled triangle. True/**False**/Uncertain
   (4) The triangle cannot be both equilateral and right-angled at the same time.
       True/False/Uncertain

(5) The triangle can be both right-angled and isosceles at the same time.

True/False/Uncertain

Marking scheme: Full mark 1' for each; if choose wrongly or choose 'uncertain', assign 0'.

4. State whether the following statements are True or False (Full mark: 1' × 4 = 4').

(1) A triangle is an acute-angled triangle when it has an acute angle.

True/**False**/Uncertain

(2) The sum of the angles of an acute-angled triangle is smaller than the sum of the angles of an obtuse-angled triangle. True/**False**/Uncertain

(3) An acute-angled triangle cannot be a scalene triangle. True/**False**/Uncertain

(4) An acute-angled triangle has no symmetrical axis. True/**False**/Uncertain

Marking scheme: Full mark 1' for each; if choose wrongly or choose 'uncertain', assign 0'.

5. State whether the following statements are True or False (Full mark: 1' × 5 = 5').

(1) A scalene triangle may have two angles equal in size.

True/**False**/Uncertain

(2) A scalene triangle cannot be an obtuse-angled triangle.

True/**False**/Uncertain

(3) A scalene triangle cannot have a right angle. True/**False**/Uncertain

(4) A scalene triangle has no symmetrical axis. **True**/False/Uncertain

(5) Triangles, except equilateral triangle are all scalene triangle.

True/**False**/Uncertain

Marking scheme: Full mark 1' for each; if choose wrongly or choose 'uncertain', assign 0'.

6. Decide whether the following statements are always true for a triangle? (Full mark: 1' × 5 = 5')

(1) If it has a right angle, then all its sides are of different length.

Yes/**No**/Uncertain

(2) If two of its sides are equal in length, then none of its angles is a right angle.

Yes/No/Uncertain

(3) If at least two of its angles are equal in size, then all its three sides are equal in length. Yes/**No**/Uncertain

(4) If it has a right angle, then at most two of its sides can be equal in length.

Yes/No/Uncertain

(5) If one of its angles is obtuse, then all its sides are different in length.

Yes/**No**/Uncertain

Marking scheme: Full mark 1' for each; if choose wrongly or choose 'uncertain', assign 0'.

7. Decide whether the following properties are always true for every isosceles triangle (Full mark: 1' × 4 = 4').

    (1) One of its sides is always shorter than the other sides. Yes/**No**/Uncertain
    (2) All its sides are equal in length.   Yes/**No**/Uncertain
    (3) None of its angles is an obtuse angle.   Yes/**No**/Uncertain
    (4) Two of its angles are equal in size.   **Yes**/No/Uncertain
    Marking scheme: Full mark 1' for each; if choose wrongly or choose 'uncertain', assign 0'.

8. State whether the following statements are True or False (Full mark: 1' × 5 = 5').

    (1) Every equilateral triangle is an isosceles triangle because every isosceles triangle has all the properties of an equilateral triangle.
                                                  True/**False**/Uncertain
    (2) Every equilateral triangle is an isosceles triangle because every equilateral triangle has all the properties of an isosceles triangle.
                                                  True/False/Uncertain
    (3) Every isosceles triangle is an equilateral triangle because every property of an isosceles triangle is a property of an equilateral triangle.
                                                  True/**False**/Uncertain
    (4) An equilateral triangle is not an isosceles triangle because they have different properties.   **True**/False/Uncertain
    (5) Every isosceles triangle is not an equilateral triangle because they looked different.
                                                  True/**False**/Uncertain
    Marking scheme: Full mark 1' for each; if choose wrongly or choose 'uncertain', assign 0'.

9. State whether the following statements are True or False (Full mark: 1' × 4 = 4').

    (1) The intersection points of a triangle's heights, medians, and angle bisectors are all inside the triangle.   True/**False**/Uncertain
    (2) Triangle can be divided into five categories: acute-angled triangle, right-angled triangle, obtuse-angled triangle, isosceles triangle, and equilateral triangle.
                                                  True/**False**/Uncertain
    (3) The exterior angle of a triangle is equal to the sum of two interior angles of the triangle.   True/**False**/Uncertain
    (4) The exterior angle of a triangle and its adjacent interior angle are adjacent supplementary angles.   **True**/False/Uncertain

Marking scheme: Full mark 1' for each; if choose wrongly or choose 'uncertain', assign 0'.

10. Given two fixed points A and B, how many isosceles right-angled triangles of different positions can you draw?   (C)

    (A) 2   (B) 4   (C) 6   (D) 8
    Marking scheme: Full mark 2' for correct answer 'C'; other answers, assign 0'.

11. Which of the following groups of numbers can be the measure of the three sides in a triangle?   (AD)

    (A) 6, 7, 3   (B) 3, 4, 7   (C) 2, 9, 5   (D) 4, 7, 4   (E) 13, 14, 15
    Marking scheme: Full mark 2' for correct answer 'AD'; 1' for partly correct answer.

12. Given the length of two sides of an isosceles triangle 11 cm and 5 cm, what is the perimeter of the triangle?   (B)

    (A) 21 cm   (B) 27 cm   (C) 21 cm or 27 cm   (D) Uncertain
    Marking scheme: Full mark 2' for correct answer 'B'; other answers, assign 0'.

13. Which of the following points has the same distance to the three sides of a triangle? (A)

    (A) The intersection point of the angular bisectors of its internal angles.
    (B) The intersection point of the heights of the triangle.
    (C) The intersection point of the median.
    (D) The intersection point of the perpendicular bisectors of its three sides.
    Marking scheme: Full mark 4', 1' for each option depending on whether it should be chose.

14. Here are two statements:

    S1: △ABC has three sides of the same length.
    S2: In △ABC, ∠ B and ∠ C have the same measure.
    Which is correct? (B)
    (A) S1 and S2 cannot both be true.   (C) If S2 is true, then S1 is true.
    (B) If S1 is true, then S2 is true.   (D) If S1 is false, then S2 is false.
    Marking scheme: Full mark 2' for correct answer 'B'; 1' for partly correct answer.

15. If the two altitudes of a triangle intersect outside the triangle, then what kind of triangle is that? Draw an example below and check your conclusion (Full mark: 2').

    Test objective: Obtuse-angled triangle; heights of a triangle;
    Marks awarded:
                1' . . . . . . Answer: Obtuse-angled triangle
                1' . . . . . . Correct example

16. In ΔABC, E and F are the midpoints of AB and AC respectively.

    If EF=1.5, then BC=_____. (Full mark: 2')
    Test objective: Property of midline in a triangle;

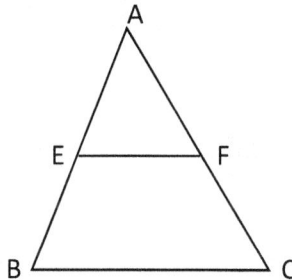

Marks awarded:
        1' . . . . . . BC = 3
        1' . . . . . . Reasoning: The midline in a triangle connecting mid-
points of its two sides is parallel to a third side and its length is one half
of the length of the third side.

17. In ΔABC, BO bisects ∠ ABC, CO bisects ∠ ACB, MN // BC, MN
    passes by point O, AB = 12, and AC = 8, then the perimeter of ΔAMN
    = _____. (Full mark: 3')

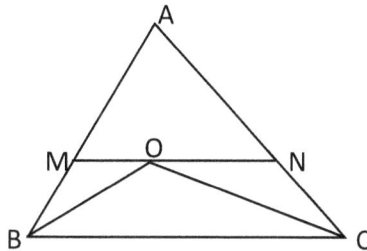

Test objective: Properties of bisector of an angle, parallel lines & isosceles
    triangle; definition of circumference.
Marks awarded:
        1' . . . . . . perimeter = 20
        2' . . . . . . Reasoning (Mentioned any two of the three: property
of bisector of an angle; property of parallel lines; property of isosceles
triangle.)

18. In the figure, ΔABC is an isosceles triangle, AB=AC, D is a point inside the triangle and DB=DC. If∠ ABD=15°, ∠ BCD=35°. Find the value of ∠ BAC and ∠ BDC. (Full mark: 4')

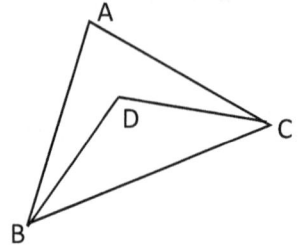

Test objective: Properties of isosceles triangle; angle sum of a triangle.
Marks awarded:

        1'......∠ ABC = ∠ ACB, ∠ DBC = ∠ DCB
               (Use the same property of isosceles triangle)
        1'...... Angle sum of a triangle 180°
        1'......∠ BAC = 80°
        1'......∠ BDC = 110°

19. In the figure, P is a point inside equilateral triangle ABC, PA=6, PB=8, PC=10. Counter-clockwise ΔPAC about Point A to the position of ΔP'AB. Find the length of PP' (Full mark: 4').

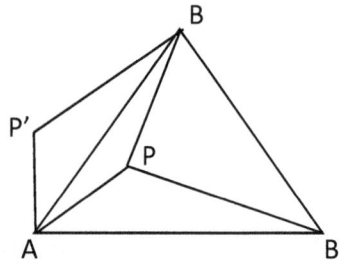

Test objective: Property of rotation, properties of an equilateral triangle
Marks awarded:

        1'......∠ PAP' = 60°;
        1'...... P'A = PA = 6;
        1'...... Equilateral triangle PP'A;
        1'...... PP' = 6.

20. In the figure, in RtΔABC, ∠ ACB = 90°, AD is the median to BC, and AD = BC, BE ⊥ AD, prove that AD = 4DE. (Full mark: 7')

(Item deleted: mathematics teacher intervened)

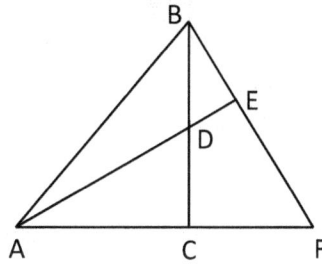

Test objective: Properties of RtΔABC, definition of median, property of a median of the hypotenuse in an RtΔABC.

Marks awarded:

Method 1: 1...... Median: BD = CD
           1...... CD = AD/2
           1...... RtΔABC or ∠ ACB = 90°
           1......∠ DAC = 30°
           1......∠ DBE = 30°
           1...... DE = BD/2
           1...... DE = AD/4

Method 2: 1...... Subline CM, M is the midpoint of AD
           1...... CM = AD/2
           1...... Median: CD = BD
           1...... CM = CD
           1...... Equilateral triangle CND, ∠ CDA = 60°
           1...... DE = BD/2
           1...... DE = AD/4

Method 3: 1...... Subline CN, so that CN ⊥ BF or EN = BE
          1...... Midline DE in ΔBCG, DE = CG/2
          3......ΔABC ≅ ΔABC (need several conditions, angles equal
   or sides equal)
          1...... CG = CD = BC/2 = AD/2
          1...... DE = AD/4

Method 4: 1...... Subline GH, so that GH is the midline of ΔADC. G is the
   midpoint
of AD and H is the midpoint of DC.
          1...... Midline GH in ΔDAC, GH ⊥ DC
3...... ΔDGH ≅ ΔDBE (need several conditions, angles equal or sides
   equal)
          1...... DH = CD/2 = BC/4 = AD/4
          1...... DE = DH

Traditional Paper-and-Pencil Task (P&P Task) (Quadrilateral)

—Answers and Marking Scheme

1. Which of the following figures are parallelograms? Please write the corresponding letters on the line below. (Full mark 2')

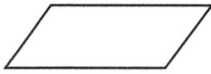  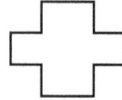 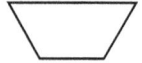

A   B   C   D

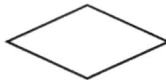 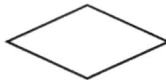  

E   F   G   H

Parallelogram: *AFGH*   .

Marking scheme: Full mark 2'; 1' for partly correct answer. The correct answer is provided on the line above.

2. State whether each of the following statements is True or False.

(Full mark: 1' × 4 = 4')

(A) All properties of rectangles are properties of all squares.

True/False/Uncertain

(B) All properties of squares are properties of all rectangles.

True/**False**/Uncertain

(C) All properties of rectangles are properties of all parallelograms.

True/**False**/Uncertain

(D) All properties of squares are properties of all parallelograms.

True/**False**/Uncertain

Marking scheme: Full mark 1' for each; if choose wrongly or choose 'uncertain', assign 0'. Answers are bold, hereinafter the same.

3. Which of the following properties do all rectangles have that some parallelograms do not have?   (B)

(A) Opposite sides equal
(B) Diagonals equal
(C) Opposite sides parallel
(D) Opposite angles equal

Marking scheme: Full mark 4'. 1' for each option, depending on whether it should be chosen or not.

4. In a rectangle ABCD, AC and BD are the diagonals.

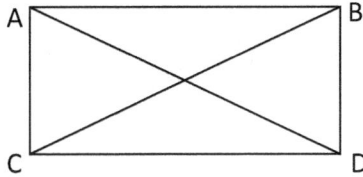

Which of (A)–(D) is true in *every* rectangle?   (ABCD)

(A) There are four right angles   (B) There are four sides.

(C) The diagonals have the same length.   (D) The opposite sides have the same length.

Marking scheme: Full mark 4'. 1' for each option, depending on whether it should be chosen or not.

5. A rhombus is a four-sided figure with all sides of the same length. Here are three examples:

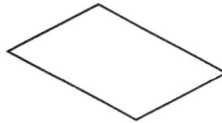

Which of (A)–(D) is *not* true in every rhombus?   (BCD)

(A) The two diagonals have the same length.

(B) Each diagonal bisects two angles of the rhombus.

(C) The two diagonals are perpendicular.

(D) The opposite angles have the same measure.

Marking scheme: Full mark 4'. 1' for each option, depending on whether it should be chosen or not.

6. A particular shape has the following properties:

It has four sides.

Adjacent sides are not equal in length.

Diagonals are equal in length.

Adjacent sides are perpendicular

This shape must be a   (B)

(A) Square   (B) Rectangle

(C) Rhombus   (D) Right-angled triangle

Marking scheme: Full mark 4' for correct answer 'B'. Assign 3' for 'A' since square satisfies three of the four properties; assign 1' for option 'C' since it satisfies the first property. Assign 2' for 'BD' and 0' for 'D'.

7. Which of the following statements is always true for any four-sides closed figure?   (B)

    (A) If its diagonals bisect each other, then all its sides are equal in length.
    (B) If its diagonals bisect each other, then its opposite sides are equal in length.
    (C) If its opposite sides are parallel, then none of its angles is a right angle.
    (D) If its opposite sides are equal in length, then the adjacent sides are perpendicular.
    (E) If its diagonals are equals are equal in length, then the adjacent sides are perpendicular.
    Marking scheme: Full mark 5'; 1' for each option, depending on whether it should be chosen.

8. State whether each of the following statements is True or False (Full mark: 1' × 4 = 4').

    (1) A quadrilateral is a parallelogram when it has one pair of opposite sides parallel and the other pair of opposite sides equal in length.
    True/**False**/Uncertain
    (2) A quadrilateral is a parallelogram when it has bisecting diagonals.
                                                           True/False/Uncertain
    (3) A quadrilateral is a parallelogram when it has one pair of opposite sides parallel and one pair of opposite angles equal.   **True**/False/Uncertain
    (4) A quadrilateral is a parallelogram when its two pairs of opposite sides are respectively equal in length.   **True**/False/Uncertain
    Marking scheme: Full mark 4', 1' for each; if choose wrongly or choose 'uncertain', assign 0'. Answers are bold, hereinafter the same.

9. State whether each of the following statements is True or False (Full mark: 1' × 5 = 5').

    (1) A rhombus is not a parallelogram because they looked different.
    True/**False**/Uncertain
    (2) A parallelogram cannot be a rhombus because they have different properties.
                                                           True/**False**/Uncertain
    (3) Every rhombus is a parallelogram because every property of a parallelogram is a property of a rhombus.   **True**/False/Uncertain
    (4) Every parallelogram is a rhombus because every property of a parallelogram is a property of a rhombus.   True/**False**/Uncertain
    (5) Parallelograms and rhombuses are the same because they are just different names for the same shape.   True/**False**/Uncertain
    Marking scheme: Full mark 5', 1' for each; if choose wrongly or choose 'uncertain', assign 0'.

10. Here are three statements about a quadrilateral:

    S1: The diagonals bisect each other.
    S2: All its sides are equal in length.
    S3: Opposite sides are parallel.
    Which of the following is true   (C)
    (A) S1 implies S2 which implies S3.   (B) S1 implies S3 which implies S2.
    (C) S2 implies S3 which implies S1.   (D) S3 implies S1 which implies S2.
    (E) S3 implies S2 which implies S1.
    Marking scheme: Full mark 3' because this item tests three relationships
        between the statements: S1 and S2, S1 and S3, and S2 and S3; Assign
        1' for option 'A' since the relationship between S2 and S3 is correct.
        Assign 1' for option 'D' since the relationship between S1 and S3 is cor-
        rect. Assign 1' for partly correct answers like 'ACD', 'BCD', 'CE', and
        'BCE'. Assign 0' for answer 'B' since two of the three relationships are
        judged wrongly and the third one is not mentioned.

11. Which of the following quadrilateral could *not* be a trapezium?   (A)

    (A) A quadrilateral with two pairs of equal sides.
    (B) A quadrilateral with one pair of opposite sides equal in length.
    (C) A quadrilateral with two right angles.
    (D) A quadrilateral with two equal diagonals.
    (E) A quadrilateral with two perpendicular diagonals.
    Marking scheme: Full mark 5' for correct answer 'A'; 1' for each option
        depending on whether it should be chosen.

12. In the figure on the right, the lengths of the sides are shown.

    Which of the following guarantees the figure is a rectangle?   (C)
    (A) The opposite sides are congruent.
    (B) The opposite angles are congruent.
    (C) The angles are right angles.
    (D) The opposite sides are parallel.
    (E) Hard to decide.

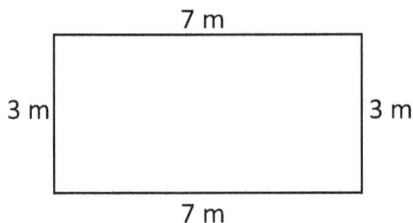

Marking scheme: Full mark 4'; 1' for each option depending on whether it
    should be chosen. Assign 0' for 'E'.

13. Which of the following properties are correct about a square?   (ABD)

    (A) Square has four symmetrical axes.
    (B) A rhombus with equal diagonals is a square.
    (C) Square is a special kind of isosceles trapezium.
    (D) Square is a special kind of rectangle.
    Marking scheme: Full mark 4'; 1' for each option depending on whether it
    should be chosen.

14. In the figure, ABCD is a parallelogram, diagonals AC and BD intersects at
    point O. If AC = 12, BD = 10, AB = m, find the range of m: $1 < m < 11$.

    (Full mark 2')
    Marking scheme: Full mark 2'. Assign 1' for half of the inequality correct and
       0' for totally wrong answer. Minor details such as $1 < x < 11$ can be ignore.

15. Given the property of a quadrilateral, please write down the name and show
    the corresponding figure of the specific quadrilateral which satisfies the prop-
    erty. (Full mark 3' × 2 = 6')

    (1) The two diagonals are perpendicular.
        Correct answer: square, rhombus, and one of the following special
        cases: kite, special trapezium (with perpendicular diagonals), or special
        quadrilateral (with perpendicular diagonals).
        Marking scheme: Full mark 3'. 1' for square, 1' for rhombus, and 1' for
        one of the special cases, mentioned either in name or in figure. (In China,
        kite is not required in secondary textbook but teachers may mention it
        in class. The basic consideration of the scoring is that students should
        know the property "perpendicular diagonals" does not limit to square
        and rhombus. They should have a sense to consider other quadrilaterals.)
    (2) The two diagonals are equal in length.
        Correct answer: square, rectangle, and isosceles trapezium.
        Marking scheme: Full mark 3'. 1' for square, 1' for rhombus, and 1' for
        isosceles trapezium.

16. Consider each of the special quadrilaterals in turn. (Full mark 3' × 2 = 6')

    (1) Does the diagonal divide the quadrilateral into two halves equal in area?
        Correct answer: Not always. For example, the diagonals of trapezium
        cannot divide it into two halves equal in area. (Provide a counter-example)
        Marking scheme: Full mark: 3'. 1' for answer; 2' for reasoning, 1' for
        partly correct reasoning.
    (2) Is any diagonal a line of symmetry?
        Correct answer: No. For square and rhombus, their diagonals are lines
        of symmetry, but for rectangle and trapezium, the diagonals are not their
        symmetry axes.
        Marking scheme: Full mark: 3'. 1' for answer; 2' for reasoning, 1' for
        partly correct reasoning.

17. In rectangle ABCD, the two diagonals intersect at point N. If ∠ABD = 74°, find the values for ∠DNC and ∠ACB. (Full mark: 4')

Marking scheme: ∠DNC =32° . . . . . . 1'

∠ACB =16° . . . . . . 1'

Reasoning:

Mention any two of the three properties: Rectangle 90°; Bisecting diagonal; Angle sum 180° . . . . . . 2'

(Students are quite familiar with problems of this type. They may slip small steps in the answer. Thus, it is acceptable that if they mentioned only two of the three properties. If they mentioned only one, assign 1' for reasoning; or if none of the three is mentioned, assign 0' for reasoning.)

18. Determine the value of x and y in the rhombus. (Full mark: 4')

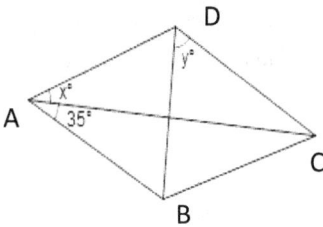

Marking scheme:

M1: Answer: . . . . . . 2' (1' for x, 1' for y)

Reasoning: 2'. Mention any of the two properties below:

Rhombus, diagonal bisecting interior angles. . . . . . 1'

Perpendicular diagonals. . . . . . 1'

Angle sum 180° . . . . . . 1'

M2: Answer: . . . . . . 2' (1' for x, 1' for y)

Reasoning: 2', mention any of the two properties below.

Rhombus, diagonal bisecting interior angles. . . . . . 1'

Equal adjacent sides, two angles equal. . . . . . 1'

Parallel sides, two angles equal. . . . . . 1'

19. In a parallelogram, the difference of two adjacent angles is 23°. Find the value of all the four angles in the parallelogram. (Full mark 3')

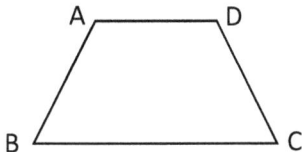

Marking scheme:

Sum of two adjacent angles 180° . . . . . . 1'

Answers: . . . . . . 2' (1' for each)

20. Given Trapezium ABCD, AD//BC, ∠ B and ∠ D are supplementary. Decide whether ABCD is an isosceles trapezium. Give you reasons briefly (Full mark: 5')

   Marking scheme: (A formal proof should include drawing a subline and convey the problem into proving two sides equal in a triangle. The definition of isosceles trapezium is "the two non-parallel sides equal" instead of "two basilar angles equal". The subline step is required for strict proof.)

   Subline DE // AB...... 1'
   AB = DE, ∠ DEC = ∠ B...... 1'
   ∠ B + ∠ D = ∠ D = ∠ C = 180°...... 1'
   ∠ B = ∠ C = ∠ DEC => DE = DC...... 1'
   DC = AB => Isosceles trapezium ABCD...... 1'

21. In ▱ABCD, E and F are the middle points of segments AB and CD respectively. BD is the diagonal of ▱ABCD and AG//DB. Given that quadrilateral BEDF is a rhombus, what kind of quadrilateral is AGBD? Please explain your answer briefly.

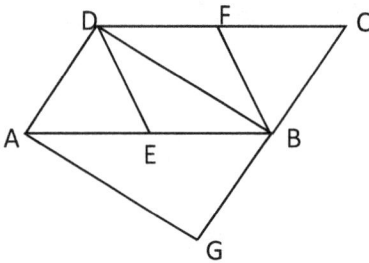

Marking scheme:
M1:   ▱ABCD => AD // BC...... 1'
      ▱AGBD...... 1'
      Rhombus DEBF => DE = BE...... 1'
      Midpoint E => AE = BE = AB/2...... 1
      '∠ DAE = ∠ ADE, ∠ EDB= ∠ EBD, ∠ DAE + ∠ ADE + ∠ EDB + ∠ EBD = 180°... 1'
      ∠ ADB = 90°...... 1'
      Rectangle AGBD...... 1'
M2: ▱ABCD => AD // BC...... 1'
      ▱AGBD...... 1'
      AD = BG, ∠ DAE = ∠ EBG...... 1'
      AE = BE, △ADE ≅ △BEG...... 1'
      ∠ DEA = ∠ BEG => E, D, G collinear...... 1'
      DG = AB...... 1'
      Rectangle AGBD...... 1'

Appendix E

Attitudes Toward Concept Mapping Questionnaire (ATCMQ)

Name: _____ Class: _____
Class Register No.: _____ Gender: _____ Age: _____.

INSTRUCTIONS:

The purpose of this study is to find out your attitudes toward using concept maps in mathematics. It is NOT a test, and there are no "right" or "wrong" answers. **A concept map** refers to a diagram of linked concepts showing overall structure and details. It includes concepts (in boxes or circles), links (with directional arrow from one concept to another), and linking phrases (describing the relationships between linked concepts).

Please read each statement carefully and indicate your agreement to it by circling ONE of the six numbers below.

1 = Strongly Agree (SA)
2 = Agree (A)
3 = Slightly Agree (LA)
4 = Slightly Disagree (LD)
5 = Disagree (D)
6 = Strongly Disagree (SD)

No.	Item	Response					
		SA	A	LA	LD	D	SD
1	I can write statements from a given concept map.	1	2	3	4	5	6
2	I can read the hierarchy among concepts in a given map.	1	2	3	4	5	6
3	Concept mapping is helpful for understanding mathematics concepts.	1	2	3	4	5	6
4	Concept mapping is time-consuming.	1	2	3	4	5	6

(*Continued*)

No.	Item	Response					
		SA	*A*	*LA*	*LD*	*D*	*SD*
5	I can judge whether the linking phrases in a given map are accurate.	1	2	3	4	5	6
6	Given more time, I can add more propositions to my concept map.	1	2	3	4	5	6
7	I can distinguish a good map from a bad one.	1	2	3	4	5	6
8	Concept map can reflect what I understand about the concepts.	1	2	3	4	5	6
9	Concept mapping is easy to construct.	1	2	3	4	5	6
10	It is fair to judge our conceptual achievement according to our performance in a concept map.	1	2	3	4	5	6
11	I am not sure whether my concept map is good.	1	2	3	4	5	6
12	I will try to use concept map in my further study.	1	2	3	4	5	6
13	I find concept mapping boring.	1	2	3	4	5	6
14	I hope our teacher can use concept map to teach us mathematical concepts.	1	2	3	4	5	6
15	I find it difficult to add accurate linking phrases.	1	2	3	4	5	6
16	I like spending time on concept mapping.	1	2	3	4	5	6
17	Using concept map, one can clearly describe relationships between mathematical concepts.	1	2	3	4	5	6
18	I'd like to use concept map in mathematics.	1	2	3	4	5	6
19	I find it easy to tell the hierarchy among a list of given concepts.	1	2	3	4	5	6
20	Concept Mapping is interesting to me.	1	2	3	4	5	6
21	Concept mapping is a waste of time.	1	2	3	4	5	6
22	I am good at concept mapping.	1	2	3	4	5	6
23	I feel excited when I heard the word 'concept map'.	1	2	3	4	5	6
24	I have no idea about how to read a concept map.	1	2	3	4	5	6
25	I can learn mathematics better if my teacher uses concept mapping for teaching.	1	2	3	4	5	6
26	I feel anxious when I am asked to construct a concept map.	1	2	3	4	5	6
27	I can come up with new ideas when doing concept mapping.	1	2	3	4	5	6
28	I feel lost when trying to construct a concept map.	1	2	3	4	5	6

No.	Item	Response					
		SA	A	LA	LD	D	SD
29	After doing a concept map, I can see more clearly how the concepts are related.	1	2	3	4	5	6
30	I would like to have more concept mapping activities in my mathematics lessons.	1	2	3	4	5	6
31	I have to think harder when doing concept mapping.	1	2	3	4	5	6
32	I believe that concept mapping is useful.	1	2	3	4	5	6
33	I don't think I can do well on concept mapping.	1	2	3	4	5	6
34	I find concept mapping challenging.	1	2	3	4	5	6
35	I am sure I can construct better concept maps if I get more familiar with how to construct them.	1	2	3	4	5	6

*** **Check that you have answered every question. Thank you.** ***

References

Afamasaga-Fuata'I, K. (2006) 'Developing a more conceptual understanding of matrices and systems of linear equations through concept mapping and Vee diagrams', *Focus on Learning Problems in Mathematics*, 28 (3 and 4), pp. 58–89.

Afamasaga-Fuata'I, K. (2008) 'Students' conceptual understanding and critical thinking: A case for concept maps and vee-diagrams in mathematics problem solving', *The Australian Mathematics Teacher*, 64 (2), pp. 8–17.

Afamasaga-Fuata'I, K. (2009a) 'Analysing the "measurement" strand using concept maps and vee diagrams', in K. Afamasaga-Fuata'I (ed), *Concept mapping in mathematics: Research into practice*. New York, NY: Springer, pp. 19–46.

Afamasaga-Fuata'I, K. (2009b) 'Using concept maps and vee diagrams to analyse the "fractions" strand in primary mathematics', in K. Afamasaga-Fuata'I (ed), *Concept mapping in mathematics: Research into practice*. New York, NY: Springer, pp. 59–86.

Ahlberg, M. (2004) 'Varieties of concept mapping', in A. J. Cañas, J. D. Novak and F. M. Gonzalez (eds), *Concept maps: Theory, methodology, technology*. Proceedings of the First Conference on Concept Mapping. Available at http://cmc.ihmc.us/CMC2004Pro grama.html (downloaded: 25 March 2008)

Al-Mutawah, M. A., Thomas, R., Eid, A., Mahmoud, E. Y., and Fateel, M. J. (2019) 'Conceptual understanding, procedural knowledge and problem-solving skills in mathematics: High school graduates work analysis and standpoints', *International Journal of Education and Practice*, 7 (3), pp. 258–273.

Alonso-Tapia, J. (2002) 'Knowledge assessment and conceptual understanding', in M. Limon and L. Mason (eds), *Reconsidering conceptual change: Issues in theory and practice*. Hingham, MA: Kluwer Academic Publishers, pp. 389–413.

Anderson, T. H., and Huang, S.-C. C. (1989) *On using concept maps to assess the comprehension effects of reading expository text* (Technical Report No. 483). Urbana-Champaign: Centre for the Studying of Reading, University of Illinois at Urbana-Champaign. ERIC Document Reproduction Service No. ED 310 368

Atlgan, H., Demir, E. K., Ogretmen, T., and Baoku, T. O. (2020) 'The use of open-ended questions in large-scale tests for selection: Generalizability and dependability', *International Journal of Progressive Education*, 16 (5), pp. 216–227.

Baroody, A. J., and Bartels, B. H. (2000) 'Using concept maps to link mathematical ideas', *Mathematics Teachers in the Middle School*, 5 (9), pp. 604–609.

Bartels, B. J. (1995) 'Examining and promoting mathematical connections with concept mapping', unpublished doctoral dissertation, University of Illinois at Urbana-Champaign, USA.

Bell, A. W., Costello, J., and Kuchemann, D. (1983) *A review of research in mathematics education: Part A*. Windsor, Berks: NFER-Nelson.

Bell, A. W., Küchemann, D., and Costello, J. (1983) *Research on learning and teaching*. Windsor, Berks: NFER-Nelson.

Bereiter, C., and Scardamalia, M. (1998) 'Beyond Bloom's taxonomy: Rethinking knowledge for the knowledge age', in A. Hargreaves, A. Lieberman, M. Fullan and D. Hopkins (eds), *International handbook of educational change*. London: Kluwer Academic Publishers, pp. 675–692.

Bolte, L. A. (1999) 'Using concept maps and interpretive essays for assessment in mathematics', *School Science and Mathematics*, 99 (1), pp. 19–29.

Boxtel, C. V., Linden, J. V., Roelofs, E., and Erkens, G. (2002) 'Collaborative concept mapping: Provoking and supporting meaningful discourse', *Theory into Practice*, 41 (1), pp. 40–46.

Bransford, J. D., Brown, A. L., and Cocking, R. R. (eds) (1999) *How people learn: Brain, mind, experience, and school*. Washington, DC: National Academy Press.

Castellón, L. B., Burr, L. G., and Kitchen, R. S. (2011) 'English language learners' conceptual understanding of fractions', in K. Téllez, J. Moschkovich and M. Civil (eds), *Latinos/as and mathematics education: Research on learning and teaching in classrooms and communities*. Charlotte, NC: Information Age.

Ceran, S. A., and Ates, S. (2020) 'Conceptual understanding levels of students with different cognitive styles: An evaluation in terms of different measurement techniques', *Eurasian Journal of Educational Research*, 8 (8), pp. 149–178.

Chinnappan, M., and Lawson, M. J. (2005) 'A framework for analysis of teachers' geometric content knowledge and geometric knowledge for teaching', *Journal of Mathematics Teacher Education*, 8, pp. 197–221.

Chiu, M. H., Guo, C. J., and Treagust, D. F. (2007) 'Assessing students' conceptual understanding in science: An introduction about a national project in Taiwan', *International Journal of Science Education*, 29 (4), pp. 379–390.

Cliburn, J. W. (1986) 'Using concept maps to sequence instructional materials'. *Journal of College Science Teaching*, 15 (4), pp. 377–379.

Crooks, N., and Alibali, M. W. (2014) 'Defining and measuring conceptual knowledge in mathematics', *Developmental Review*, 34 (4), pp. 344–377.

de Vries, E., Lund, K., and Baker, M. J. (2002) 'Computer-mediated epistemic dialogue: Explanation and argumentation as vehicles for understanding scientific notions', *Journal of the Learning Sciences*, 11 (1), pp. 63–103.

Dochy, F. (1994) 'Prior knowledge and learning', in T. Husén and T. Postlethwaite (eds), *The International Encyclopedia of Education*. 2nd edn. Oxford: Pergamon, pp. 4698–4702.

Domin, D., and Bodner, G. (2012) 'Using students' representations constructed during problem solving to infer conceptual understanding', *Journal of Chemical Education*, 89 (7), pp. 837–843.

Drake, B. M., and Amspaugh, L. B. (1994) 'What writing reveals in mathematics', *Focusing on Learning Problems in Mathematics*, 16 (3), pp. 43–50.

Edwards, B. W. (1993) 'The effects of using computer-based organizational software for generating mathematics-related concept maps', unpublished doctoral dissertation, Southern Illinois University as Carbondale, IL.

Edwards, J., and Fraser, K. (1983) 'Concept maps as reflectors of conceptual understanding' *Research in Science Education*, 13, pp. 19–26.

Freeman, L. A. (2004) 'The power and benefits of concept mapping: Measuring use, usefulness, ease of use, and satisfaction', in A. J. Cañas, J. D. Novak and F. M. Gonzalez (eds), *Concept maps: Theory, methodology, technology*, Proceedings of the First International Conference on Concept Mapping. Available at http://cmc.ihmc.us/CMC2004Programa.html (downloaded: 25 March 2008).

Gao, H., Shen, E., Losh, S., and Turner, J. (2007) 'A review of studies on collaborative concept mapping: What have we learned about the technique and what is next?', *Journal of Interactive Learning Research*, 18 (4), pp. 479–492.

Gu, L., Huang, R., and Marton, F. (2004) 'Teaching with variation: A Chinese way of promoting effective mathematics learning', in L. Fan, N. Y. Wong, J. Cai and S. Li (eds), *How Chinese learn mathematics: Perspectives from insiders*. River Edge, NJ: World Scientific Publishing Co., pp. 309–347.

Hair, J. F., Jr., Anderson, R. E., Tatham, R. L., and Black, W. C. (1998) *Multivariate data analysis*. Upper Saddle River, NJ: Prentice-Hall.

Hang, K. H. (1984) 'The Van Hiele levels of geometric thought of secondary school and junior college students', unpublished master's thesis. National Institute of Education, Nanyang Technological University, Singapore.

Harnisch, D. L., Sato, T., Zheng, P., Yamagi, S., and Connell, M. (1994) 'Concept mapping approach and its applications in instruction and assessment', presented at the meeting of the American Educational Research Association, New Orleans, LA.

Hasemann, K., and Mansfield, H. (1995) 'Concept mapping in research on mathematical knowledge development: Background, methods, findings and conclusions', *Educational Studies in Mathematics*, 29, pp. 45–72.

Hauer, K. E., Boscardin, C., Brenner, J. M., van Schaik, S. M., and Papp, K. K. (2020) 'Twelve tips for assessing medical knowledge with open-ended questions: Designing constructed response examinations in medical education', *Medical Teacher*, 42 (8), pp. 880–885.

Heinze-Fry, J. A., and Novak, J. D. (1990) 'Concept mapping brings long-term movement toward meaningful learning', *Science Education*, 74 (4), pp. 461–472.

Herl, H. E., O'Neil, H. F., Chung, G. K. W. K., and Schacter, J. (1999) 'Reliability and validity of a computer-based knowledge mapping system to measure content understanding', *Computer in Human Behavior*, 15, pp. 315–333.

Hiebert, J., and Lefevre, P. (1986) 'Conceptual and procedural knowledge in mathematics: An introductory analysis', in J. Hiebert (ed), *Conceptual and procedural knowledge: The case of mathematics*. Hillsdale, NJ: Lawrence Erlbaum Associates, pp. 1–17.

Horton, P. B., McConney, A. A., Gallo, M., Woods, A. L., Senn, G. J., and Hamelin, D. (1993) 'An investigation of the effectiveness of concept mapping as an instructional tool', *Science Education*, 77 (1), pp. 95–111.

Hough, S., O'Rode, N., and Terman, N. (2007) 'Using concept maps to assess change in teachers' understanding of algebra: A respectful approach', *Journal of Mathematics Teacher Education*, 10, pp. 23–41.

Hoz, R., Tomer, Y., and Tamir, P. (1990) 'The relations between disciplinary and pedagogical knowledge and the length of teaching experience of biology and geography teachers', *Journal of Research in Science Teaching*, 27 (10), pp. 973–985.

Jin, H. (2007) 'On the internal networks of middle school students' mathematics knowledge: Elementary function [In Chinese 中学生函数知识网络的调查研究]', unpublished master's thesis, Nanjing Normal University, Nanjing, China.

Jin, H., Lu, J., and Zhong, Z. (2015) 'Exploration into Chinese mathematics teachers' perceptions of concept map', in L. Fan, N. Y. Wong, J. Cai and S. Li (eds.), *How Chinese teach mathematics: Perspectives from insiders*. River Edge, NJ: World Scientific Publishing Co., pp. 591–618.

Jin, H., and Wong, K. Y. (2010) 'A network analysis of concept maps of triangle concepts', in L. Sparrow, B. Kissane and C. Hurst (eds), *Shaping the future of mathematics education*. Proceedings of the Thirty-third Annual Conference of the Mathematics Education Research Group of Australasia Incorporated. Available at www.merga.net.au/documents/MERGA33_HaiyueandWong.pdf (downloaded: 29 April 2012).

Jin, H., and Wong, K. Y. (2011) 'Assessing conceptual understanding in mathematics with concept mapping', in B. Kaur and K. Y. Wong (eds), *Assessment in the mathematics classroom*. River Edge, NJ: World Scientific Publishing Co., pp. 67–90.

Jin, H., and Wong, K. Y. (2015) 'Mapping conceptual understanding of algebraic concepts: An exploratory investigation involving Grade 8 Chinese students', *International Journal of Science and Mathematics Education*, 13 (3), pp. 683–703.

Jin, H., and Wong, K. Y. (2021) 'Complementary measures of conceptual understanding: A case about triangle concepts', *Mathematics Education Research Journal*, online-first version, doi.org/10.1007/s13394-021-00381-y

Jonassen, D. H., Beissner, K., and Yacci, M. (1993) *Structural knowledge. Techniques for representing, conveying, and acquiring structural knowledge*. Hillsdale, NJ: Lawrence Erlbaum Associates.

Jones, I., Bisson, M., Gilmore, C., and Inglis, M. (2019) 'Measuring conceptual understanding in randomised controlled trials: Can comparative judgement help?' *British Educational Research Journal*, 45 (3), pp. 662–680.

Kane, M., and Trochim, W. M. K. (2007) *Concept mapping for planning and evaluation*. Thousand Oaks: Sage Publications.

Kankkunen, M. (2001) 'Concept mapping and Peirce's semiotic paradigm meet in the classroom environment', *Learning Environments Research*, 4, pp. 287–324.

Kilpatrick, J., Swafford, J., and Findell, B. (ed) (2001) *Adding it up: Helping children learn mathematics*. Washington, DC: National Academic Press.

Kline, P. (1994) *An easy guide to factor analysis*. London: Routledge.

Lapp, D. A., Nyman, M. A., and Berry, J. S. (2010) 'Student connections of linear algebra concepts: An analysis of concept maps'. *International Journal of Mathematical Education in Science and Technology*, 41 (1), pp. 1–8.

Lay-Dopyera, M., and Beyerbach, B. (1983) *Concept mapping for individual assessment*. Syracuse, NY: School of Education, Syracuse University.

Li, X. (2005) 'Negative influences caused by the gender imbalance of teachers' [In Chinese], *Journal of Yunnan Normal University (Humanities and Social Science)*, 6, pp. 58–60.

Linn, R. L., and Miller, M. D. (2005) *Measurement and assessment in teaching*. 9th edn. Upper Saddle River, NJ: Pearson Education.

Liu, X., and Hinchey, M. (1996) 'The internal consistency of a concept mapping scoring scheme and its effect on prediction validity' *International Journal of Science Education*, 18, pp. 921–937.

Malone, J., and Dekkers, J. (1984) 'The concept map as an aid to instruction in science and mathematics', *School Science and Mathematics*, 84 (3), pp. 220–231.

Mansfield, H., and Happs, J. (1989a) 'Using concept maps to explore students' understanding in geometry', *Thirteenth Annual Conference of the International Group for the Psychology of Mathematics Education*, Paris.

Mansfield, H., and Happs, J. (1989b) 'Difficulties in achieving long term conceptual change in Geometry', unpublished paper, Perth, Western Australia.

Mansfield, H., and Happs, J. (1991) 'Concept maps', *The Australian Mathematics Teacher*, 47 (3), pp. 30–33.

Markham, K. M., Mintzes, J. J., and Jones, M. G. (1994) 'The concept map as a research and evaluation tool: Further evidence of validity' *Journal of Research in Science Teaching*, 31 (1), pp. 91–101.

McClure, J. R., Sonak, B., and Suen, H. K. (1999) 'Concept map assessment of classroom learning: Reliability, validity, and logistical practicality', *Journal of Research in Science Teaching*, 36 (4), pp. 475–492.

Meel, D. E. (2005) 'Concept maps: A tool for assessing understanding?', in G. M. Lloyd, M. Wilson, J. L. M. Wilkins and S. l. Behm (eds), *Proceedings of the 27th annual meeting of the North American chapter of the international group for the psychology of mathematics education*. Blacksburg: Virginia Tech.

Ministry of Education, China. (2011) *National mathematics curriculum standards (compulsory education)*. Beijing: Beijing Normal University Press.

Ministry of Education, Singapore. (2019) *Mathematics syllabuses: Secondary one to four express course, normal (academic) course*. Singapore: Curriculum Planning and Development Division, Ministry of Education.

Mintzes, J. J., Wandersee, J. H., and Novak, J. D. (eds) (2000) *Assessing science understanding: A human constructivist view*. San Diego, CA: Academy Press.

Mohamed, N. B. R. A. (1993) 'Concept mapping and achievement in secondary science', unpublished master's thesis, National University of Singapore, Singapore.

Morgan, C. (2005) 'Word, definitions and concepts in discourses of mathematics, teaching and learning', *Language and Education*, 19 (2), pp. 102–116.

Moskal, B. M. (2000) 'Understanding student responses to open-ended tasks', *Mathematics Teaching in the Middle School*, 5 (8), pp. 500–505.

Mosvold, R., and Fauskanger, J. (2013) 'Teachers' beliefs about mathematical knowledge for teaching definitions' *International Electronic Journal of Mathematics Education*, 8 (23), pp. 43–59.

National Council of Teachers of Mathematics. (2000) *Principles and standards for school mathematics*. Reston, VA: Author.

Nelson, M., and Pan, A. (1995) 'Integrating the concept attainment teaching model and videodisk images', *Annual Meeting of the Midwestern Educational Research Association*, Chicago, IL.

Nesbit, J. C., and Adesope, O. (2005) 'Dynamic concept maps', in P. Kommers and G. Richards (eds), *Proceedings of world conference on educational multimedia, hypermedia and telecommunications 2005*. Chesapeake, VA: Association for the Advancement of Computing in Education, pp. 4323–4329.

Nickerson, R. S. (1989) 'New directions in educational assessment', *Educational Researcher*, 18, pp. 3–7.

Niemi, D. (1996a) 'A fraction is not a piece of pie: Assessing exceptional performance and deep understanding on elementary school mathematics', *Gifted Child Quarterly*, 40 (1), pp. 70–80.

Niemi, D. (1996b) 'Assessing conceptual understanding in mathematics: Representations, problem solutions, justifications, and explanations', *The Journal of Educational Research*, 89 (6), pp. 351–363.

Nitko, A. J., and Brookhart, S. M. (2007) *Educational assessment of students*. 5th edn. Upper Saddle River, NJ: Merrill.

Norwood, K. S., and Carter, G. (1994) 'Journal writing: An insight into students' understanding', *Teaching Children Mathematics*, 1 (3), pp. 146–148.

Novak, J. D. (1990) 'Concept mapping: A useful tool for science education', *Journal of Research in Science Teaching*, 27 (10), pp. 937–949.

Novak, J. D. (1998) *Learning, creating, and using knowledge: Concept maps as facilitative tools in schools and corporations*. Mahwah, NJ: Lawrence Erlbaum.

Novak, J. D. (2005) 'Results and implications of a 12-year longitudinal study of science concept learning', *Research in Science Education*, 35, pp. 23–40.

Novak, J. D. (2006) 'The development of the concept mapping tool and the evolution of a new model for education: Implications for mathematics education', *Focus on Learning Problems in Mathematics*, 28 (3–4), pp. 3–31.

Novak, J. D., and Cañas, A. J. (2006) 'The theory underlying concept maps and how to construct them' [Electronic Version]. *Technical Report IHMC CmapTools 2006–01, Florida Institute for Human and Machine Cognition*. Available at http://cmap.ihmc. us/Publications/ResearchPapers/TheoryUnderlying ConceptMaps.pdf. (downloaded: 1 December 2008).

Novak, J. D., and Gowin, D. B. (1984) *Learning how to learn*. Cambridge and London: Cambridge University Press.

Novak, J. D., Gowin, D. B., and Johansen, G. T. (1983) 'The use of concept mapping and knowledge Vee mapping with Junior High School science students', *Science Education*, 67 (5), pp. 625–645.

Okebukola, P. A. O. (1992) 'Attitudes of teachers towards concept mapping and vee diagramming as metalearning tools in science and mathematics', *Educational Research*, 34 (3), pp. 201–213.

Panaoura, A., Michael-Chrysanthou, P., Gagatsis, A., Elia, I., and Philippou, A. (2017) 'A structural model related to the understanding of the concept of function: Definition and problem solving', *International Journal of Science and Mathematics Education*, 15, pp. 723–740.

Panasuk, R. M. (2011) 'Taxonomy for assessing conceptual understanding in algebra using multiple representations', *College Student Journal*, 45 (2), pp. 219–232.

Pearsall, N. R., Skipper, J., and Mintzes, J. J. (1997) 'Knowledge restructuring in the life sciences: A longitudinal study of conceptual change in biology', *Science Education*, 81, pp. 193–215.

Pellegrino, J., Chudowsky, N., and Glaser, R. (eds) (2001) *Knowing what students know: The science and design of educational assessment*. Washington, DC: National Research Council, National Academy Press.

Perkins, D. (1998) 'What is understanding?', in S. Wiske (ed), *Teaching for understanding- linking research with practice*. San Francisco, CA: Jossey-Bass, pp. 39–57.

Piaget, J. (1977) *The development of thought: Equilibration of cognitive structures*. New York, NY: Viking Press.

Pine, F. (1985) *Developmental theory and clinical process*. New Haven, CT: Yale University Press.

Pirie, S. E. B., and Kieren, T. E. (1994) Growth in mathematical understanding: How can we characterize it and how can we represent it? *Educational Studies in Mathematics*, 26 (3), pp. 165–190.

Plummer, K. J. (2008) 'Analysis of the psychometric properties of two different concept-map assessment tasks's, unpublished doctoral dissertation, Brigham Young University, Provo. Available at http://lib.byu.edu/cgi-bin/TxtOnly/betsie.pl/0001/contentdm.lib. byu.edu/ETD/image/etd2281.pdf (downloaded: 21 December 2011).

Price, C., and van Jaarsveld, P. (2017) 'Using open-response tasks to reveal the conceptual understanding of learners: Learners teaching the teacher what they know about trigonometry', *African Journal of Research in Mathematics, Science and Technology Education*, 21 (8), pp. 1–17.

Rasslan, S., and Vinner, S. (1998) 'Images and definitions for the concept of increasing/decreasing function', in *Proceedings of the 22nd conference of the international group for the psychology of mathematics education*, 4. Stellenbosch, South Africa: PME, pp. 33–40.

Rice, D. C., Ryan, J. M., and Samson, S. M. (1998) 'Using concept maps to assess student learning in the science classroom: Must different methods compete?', *Journal of Research in Science Teaching*, 35 (10), pp. 1103–1127.

Roberts, L. (1999) 'Using concept maps to measure statistical understanding', *International Journal of Mathematical Education in Science and Technology*, 30 (5), pp. 707–717.

Roth, W., and Roychoudhury, A. (1993) 'The concept map as a tool for the collaborative construction of knowledge: A microanalysis of high school physics students', *Journal of Research in Science Teaching*, 30 (5), pp. 503–534.

Ruiz-Primo, M. A. (2004) 'Examining concept maps as assessment tool', in A. J. Cañas, J. D. Novak and F. M. Gonzalez (eds), *Concept maps: Theory, methodology, technology: Proceedings of the First Conference on Concept Mapping*. Available at http://cmc.ihmc.us/CMC2004Programa.html (downloaded: 25 March 2008).

Ruiz-Primo, M. A., Schultz, S. E., Li, M., and Shavelson, R. J. (2001) 'Comparison of the reliability and validity of scoring from two concept-mapping techniques', *Journal of Research in Science Teaching*, 3 (2), pp. 260–278.

Ruiz-Primo, M. A., and Shavelson, R. J. (1996) 'Problems and issues in the use of concept maps in science assessment', *Journal of Research in Science Teaching*, 33 (6), pp. 569–600.

Ruiz-Primo, M. A., Shavelson, R. J., Li, M., and Schultz, S. E. (2001) 'On the validity of cognitive interpretations of scores from alternative concept-mapping techniques', *Educational Assessment*, 7 (2), pp. 99–141.

Rumelhart, D. E. (1980) 'Schemata: The building blocks of cognition', in R. J. Spiro, B. C. Bruce and W. F. Brewer (eds), *Theoretical issues in reading comprehension: Perspectives from cognitive psychology, linguistics, artificial intelligence, and education*. Hillsdale, NJ: Lawrence Erlbaum, pp. 33–58.

Rumelhart, D. E., and Ortony, A. (1977) 'The representation of knowledge in memory', in R. C. Anderson, R. J. Spiro, and W. E. Montague (eds), *Schooling and the acquisition of knowledge*. Hillsdale, NJ: Lawrence Erlbaum, pp. 99–135.

Savander-Ranne, C., and Kolari, S. (2003) 'Promoting the conceptual understanding of engineering students through visualization', *Global Journal of Engineering Education*, 7 (2), pp. 189–199.

Schau, C., and Mattern, N. (1997) 'Use of map techniques in teaching statistics courses', *The American Statistician*, 51 (2), pp. 171–175.

Schmittau, J. (2009) 'Concept mapping as a means to develop and assess conceptual understanding in secondary mathematics teacher education', in K. Afamasaga-Fuata'I (ed), *Concept mapping in mathematics: Research into practice*. New York, NY: Springer, pp. 137–148.

Schmittau, J., and Vagliardo, J. J. (2009) 'Concept mapping as a means to develop and assess conceptual understanding in primary mathematics teacher education', in K. Afamasaga-Fuata'I (ed), *Concept mapping in mathematics: Research into practice*. New York, NY: Springer, pp. 47–58.

Schnotz, W., and Preub, A. (1997) 'Task-dependent construction of mental models as a basis for conceptual change', *European Journal of Psychology Education*, 12 (2), pp. 185–211.

Sfard, A. (1991) 'On the dual nature of mathematical conceptions: Reflections on processes and objects as different sides of the same coin', *Educational Studies in Mathematics*, 22, pp. 1–36.

Shavelson, R. J., Lang, H., and Lewin, B. (1994) *On concept maps as potential "authentic" assessments in science*. (CSE Tech. Rep. No. 388). Los Angeles: University of California, National Centre for Research on Evaluation, Standards, and Student Testing (CRESST).

Shavelson, R. J., Ruiz-Primo, M. A., and Wiley, E. D. (2005) 'Windows into the mind', *Higher Education*, 49, pp. 413–430.

Skemp, R. (1976) 'Relational understanding and instrumental understanding', *Arithmetic Teacher*, 26 (3), pp. 9–15.

Skemp, R. (1987) *The psychology of learning mathematics (expanded American edition)*. Hillsdale, NJ: Lawrence Erlbaum Associations, publishers.

Snead, D., and Snead, W. L. (2004) 'Concept mapping and science achievement of middle grade students', *Journal of Research in Childhood Education*, 18 (4), pp. 306–320.

Starr, M. L., and Krajcik, J. S. (1990) 'Concept maps as a heuristic for science curriculum development: Toward improvement in process and product', *Journal of Research in Science Teaching*, 27 (10), pp. 987–1000.

Tan, C. (2006) 'Creating thinking schools through 'Knowledge and Inquiry': The curriculum challenges for Singapore', *The Curriculum Journal*, 17 (1), pp. 89–105.

Taylor, C. H. (2008) 'Promoting mathematical understanding through open-ended tasks: Experiences of an eighth-grade gifted geometry class', dissertation, Georgia State University, USA. Available at https://scholarworks.gsu.edu/cgi/viewcontent.cgi?article=10 35andamp;context=msit_diss (downloaded: 10 March 2021).

Tirosh, D. (1999) 'Finite and infinite sets: Definitions and intuitions', *International Journal Mathematics Education Science Technology*, 30 (3), pp. 341–349.

Trafunow, D., and Sheeran, P. (1998) 'Some tests of the distinction between cognitive and affective beliefs', *Journal of Experimental Social Psychology*, 34, pp. 378–397.

Ubuz, B., and Aydın, U. (2018) 'Geometry knowledge test about triangles: Evidence on validity and reliability', *ZDM*, 50 (4), pp. 659–673.

Usiskin, Z. (1982) *Van Hiele levels and achievement in secondary school geometry* (Final report of the Cognitive Development and Achievement in Secondary School Geometry project). University of Chicago, Department of Education.

Vinner, S. (1983) 'Concept definition, concept image and the notion of function', *International Journal of Mathematical Education in Science and Technology*, 14 (3), pp. 293–305.

Vinner, S. (1992) 'The function concept as a prototype of problems in mathematics learning', in G. Harel and E. Dubinsky (eds), *The concept of function: Aspects of epistemology and pedagogy*. Washington, DC: Mathematical Association of America., pp. 195–214.

Wallace, J. D., and Mintzes, J. J. (1990) 'The concept map as a research tool: Exploring conceptual change in biology', *Journal of Research in Science Teaching*, 27 (10), pp. 1033–1052.

Wang, L. (2005) 'A survey on university students' attitudes toward concept map' [In Chinese: 大学生对概念图态度的问卷调查], *Journal of Gansu Lian He University (Natural Sciences)*, 19 (1), pp. 75–79.

Wang, X., and Tang, F. (2004) 'Concept map and its practical significance in mathematics learning' [In Chinese], *Journal of Mathematics Education*, 13 (3), pp. 16–18.

Wang, X. C., and Dwyer, F. M. (2006) 'Instructional effects of three concept mapping strategies facilitating student achievement', *International Journal of Instructional Media*, 33 (2), pp. 135–151.

Webb, N., and Romberg, T. A. (1992) 'Implications of the NCTM Standards for mathematics assessment', in T. A. Romberg (ed), *Mathematics assessment and evaluation: Imperatives for mathematics educators*. Albany, NY: State University of New York Press, pp. 37–60.

Weinberg, A., Dresen, J., and Slater, T. (2016) 'Students' understanding of algebraic notation: A semiotic systems perspective', *The Journal of Mathematical Behavior*, 43, pp. 70–88.

Weinstein, C. E., and Mayer, R. E. (1986) 'The teaching of learning strategies', in M. C. Wittrock (ed), *Handbook on research in teaching*. 3rd edn. New York, NY: Macmillan, pp. 315–327.

White, R. T., and Gunstone, R. (1992) *Probing understanding*. Philadelphia, PA: The Falmer Press.

Wiburg, K., Valdez, A., Trujillo, K., Chamberlin, B., and Stanford, T. (2016) 'Impact of math snacks games on students' conceptual understanding', *Journal of Computers in Mathematics and Science Teaching*, 35 (2), pp. 173–193.

Willerman, M., and Mac Harg, R. A. (1991) 'The concept map as an advance organizer', *Journal of Research in Science Teaching*, 28, pp. 705–711.

Williams, C. G. (1994) 'Using concept maps to determine differences in the concept image of function held by students in reform and traditional calculus classes', unpublished doctoral dissertation, University of California, Berkeley.

Williams, C. G. (1998) 'Using concept maps to assess conceptual knowledge of function', *Journal for Research in Mathematics Education*, 29 (4), pp. 414–421.

Wilson, M. (1992) 'Measuring levels of mathematical understanding', in T. A. Romberg (ed), *Mathematics assessment and evaluation: Imperatives for mathematics education*. Albany, NY: State University of New York Press., pp. 213–241.

Wong, K. Y. (1999) 'Multi-Modal approach of teaching mathematics in a technological age', *8th Southeast Asian Conference on Mathematics Education*, Ateneo de Manila University, Manila.

Yildiz, P., Atay, A., and Koza, F. S. (2020) 'Preservice middle school mathematics teachers' definitions of algebraic expression and equation', *International Journal of Contemporary Educational Research*, 7 (2), pp. 156–164.

Yin, Y., and Shavelson, R. J. (2008) 'Application of generalizability theory to concept map assessment research', *Applied Measurement in Education*, 21, pp. 273–291.

Yin, Y., Vanides, J., Ruiz-Primo, M. A., Ayala, C. C., and Shavelson, R. J. (2005) 'Comparison of two concept-mapping techniques: Implications for scoring, interpretation, and use', *Journal of Research in Science Teaching*, 42 (2), pp. 166–184.

Zhang, D., Li, S., and Tang, R. (2004) 'The "two basics": Mathematics teaching and learning in Mainland China', in L. Fan, N. Y. Wong, J. Cai and S. Li (eds), *How Chinese learn mathematics: Perspective from insiders*. Singapore: World Scientific, pp. 189–207.

Index

Page numbers in *italic* indicate a figure and page numbers in **bold** indicate a table on the corresponding page.

For Product Safety Concerns and Information please contact our EU
representative GPSR@taylorandfrancis.com
Taylor & Francis Verlag GmbH, Kaufingerstraße 24, 80331 München, Germany

www.ingramcontent.com/pod-product-compliance
Lightning Source LLC
Chambersburg PA
CBHW060258220326
41598CB00027B/4148